PORTISHEAD RADIO

"A FRIENDLY VOICE ON MANY A DARK NIGHT"

Larry Bennett

Published by New Generation Publishing in 2020

Copyright © Larry Bennett 2020

First Edition

The author asserts the moral right under the Copyright, Designs and Patents Act 1988 to be identified as the author of this work.

All Rights reserved. No part of this publication may be reproduced, stored in a retrieval system or transmitted, in any form or by any means without the prior consent of the author, nor be otherwise circulated in any form of binding or cover other than that in which it is published and without a similar condition being imposed on the subsequent purchaser.

ISBN: 978-1-80031-896-0

www.newgeneration-publishing.com

New Generation Publishing

Contents

PREFACE 1
ACKNOWLEDGEMENTS AND CREDITS 3
PROLOGUE 5
CHAPTER ONE   THE EARLY YEARS AND THE 1930S 7
CHAPTER TWO   THE WAR YEARS 57
CHAPTER THREE   POST-WAR AND THE 1950S 75
CHAPTER FIVE   THE 1970S 150
CHAPTER SIX   THE 1980S 182
CHAPTER SEVEN   THE FINAL DECADE 212
CHAPTER EIGHT   AIRCRAFT, YACHTS AND FIXED STATIONS 233
CHAPTER NINE   LIFE AT PORTISHEAD RADIO 248
ANNEX 1   RUGBY RADIO STATION by MALCOLM HANCOCK 288
ANNEX 2   LEAFIELD RADIO STATION 292
ANNEX 3   ONGAR RADIO STATION 294
ANNEX 4   DORCHESTER RADIO STATION 296
ANNEX 5   CRIGGION RADIO STATION 298
ANNEX 6   BALDOCK RADIO STATION 300
ANNEX 7   BEARLEY RADIO STATION 302
ANNEX 8   SOMERTON RADIO STATION 304
ANNEX 9   BRENT RADIO TELEPHONE TERMINAL 306
ANNEX 10 THE MORSE KEYS OF PORTISHEAD RADIO ... 308
ANNEX 11 OFFICERS IN CHARGE – BURNHAM (PORTISHEAD) RADIO 314
INDEX OF REFERENCED RADIO STATIONS 315
TABLE OF ABBREVIATIONS 315

PREFACE

This book is not intended to be a concise history of maritime radio communication in the 20th century; details of the various short-range coast radio stations in the United Kingdom are only mentioned when relevant, and detailed mention of the radio pioneers (Marconi, Hertz etc.) has been left to the many other works on this subject. However, my research into the history of the UK's Maritime Radio Service has revealed many interesting facets of the various Coast Radio Stations, and details have been mentioned when it was felt that some items were historically significant. What this book is meant to relate is the story (or indeed the rise and fall) of Portishead Radio, the U.K.'s long range maritime radio station. Originally owned by the GPO, it has survived the many changes which have taken place over the years, and was until the year 2000 a vital part of BT's international communications service.

Mention Portishead Radio to any seagoing Radio Officer of the 1950s, 1960s and 1970s, and they will all relate tales of being given turn number 47 to be worked, of struggling to maintain contact from remote areas of the world, and endeavouring to receive the latest weather bulletin in Force 9 gales. Many of the 'old-timers' will reminisce longingly over the demise of the 'Area Scheme' and the 'Pacific Watch', and even the more recently-qualified seagoing Radio Officer will surely remember the bedlam on the Morse code calling bands as ships vied with each other to obtain a turn to be worked.

The advent of high-technology communications such as electronic mail, satellite voice and telex systems, and data transmissions have all had a great impact on the services provided by Portishead Radio. Morse code became obsolete (or so we were told – in reality, there is still plenty of c.w. to be heard on the HF bands) at the end of the century, superseded by more user-friendly methods of communication. Shipowners no longer have to employ specialist Radio Officers; anyone who can use a telephone or has basic computer keyboard skills can now handle ship-to-shore communications.

Portishead Radio attempted to 'move with the times' and introduced new services that required the skills of the Radio Officer to undertake. The introduction of the aeronautical radio service in 1985 and with it the Gateway land-based facility provided much-needed business, but even these services eventually switched over to satellite communications. Indeed, the continuing development of such communications led to the ultimate demise of Portishead Radio, which made its last transmission at 1200 GMT on Sunday April 30th, 2000.

The station was a unique place to work; everyone got on with everyone else, and there was such a camaraderie between the staff that friendships founded at the station continue to this day. For many of us, it was a 'dream job' and I am sure many a tear was shed when the station finally closed.

The complete story of Portishead Radio can now be told; from the early days of 1920, when the first long range radio service was introduced at Devizes Radio, through to 2000, some 81 years (and many hundreds of employees) later.

Portishead Radio - the world's greatest maritime radio station – indeed a friendly voice on many a dark night.

Larry Bennett, March 2020

ACKNOWLEDGEMENTS AND CREDITS

I would like to thank the following who have helped in my research over the last few years;

- David Hay and Anne Archer of BT Heritage & Archives for their support and assistance, and for the use of many photographs within this book;
- Eugene Byrne of the Bristol Post archives for use of material and research;
- The publishers of Wireless World, Practical Wireless and Short-Wave magazines for allowing me to use extracts from their publications;
- The UK Hydrographic Office for use of archive material;
- Mike Wilton, Tom Ponting, Dave Hopcroft, Robin Hargreaves, Graham Powell, Roger Marshall, Brian Hill, John Lamb, and Anver Anderson for their assistance, loan of documents and photographs; and also the following colleagues who have sadly passed away during the research of this book; Brian Stewart, Phil Lewis, Don Mulholland, Ernie Meaden, Ramsay Stuart and Brian Faulkner;
- Joe McCabe for the cartoons used in this book;
- Numerous ex-Portishead Radio Officers (far too many to mention) for their stories and anecdotes;
- Numerous ex-seagoing Radio Officers for the use of their stories and reminisces;
- The Crank family and the family of William Welch of Devizes for use of their Devizes Radio Station photographs;
- Nigel Hadley for use of his information about the trawler watch;
- John Snell and Douglas Palmer for their assistance with Morse key information and photographs;
- Malcolm Hancock for his help with Rugby Radio history and background;
- Paul Hawkins for his encouragement and assistance;
- My colleagues during my time at Portishead for their company and amusement over the years – happy days indeed.

Due diligence has taken place to source the copyright holders of photographs and text used. If you are the copyright holder and you have not been credited please let me know and I will update the credits for any new edition.

Facts and dates quoted in this publication have been verified as far as possible but should any evidence of errors come to light these will be corrected for any new edition, together with any new information obtained.

PROLOGUE

PORTISHEAD RADIO – A LINK WITH HOME

Many of us 'older hacks' have, at some time or another, tramped the world in ships where cabin hot water was infrequent, a 'donkey's breakfast' bunk a possibility, H/F apparatus a rare bonus and a balanced diet a figment of the imagination. We often envied the lifestyle enjoyed by many of our luckier colleagues aboard some of the more modern and faster ships. These ocean greyhounds often ran high power, and although they nearly always quickly responded to requests for them to relay messages via Portishead Radio, the relief felt by the tramp ship operator once his messages were cleared for retransmission did not necessarily ease the feeling of frustration as the imagination saw your good Samaritan leaping towards the horizon.

For the tramp ship Radio Officer, the possession of a better class radio 'ticket' and constantly being shipped out with taxed pay in your pocket in lieu of leave did not always guarantee a better berth next time. Maintenance or operational resourcefulness such as keeping a battered radar operational or resorting to the use of the static-piercing Medium Frequency quarter kilowatt emergency spark transmitter in the South China Sea to clear urgent messages often seemed to single you out by the Depot as 'just the chap' for another cheerful rust bucket.

What has all this to do with Portishead Radio? Well, quite a lot because regardless of the size, age, type and prestige of any ship and the capability of the equipment or operator, Portishead Radio has always provided the same considerate and first-rate service for all its diverse clientele. The station's officers seemed to sense the various levels of capability on board and were consistently patient with those of us who had ancient, low sensitivity receivers and little wire in the air – something well appreciated by a host of young fresh-faced first-trippers and those 'going solo' for the first time.

Less talked about, but sincerely felt by all of us is the more intangible role played by Portishead Radio over the years and far removed from the practice of traffic handling and all the other essential services. It has provided a comforting link with home.

BBC World Service broadcasts were invaluable in telling us about the latest happenings in the U.K., but after a year or two in other zones and the tamp had finally nodded her way back into Portishead Radio's control area once again, the long-awaited direct contact between her operator and his familiar home station was as good as a tonic. You knew that, at last, your own family and friends were just a few days away.

Yes, good old Portishead Radio meant a lot to all of us – not least to the long-haul, long-away tramp ship 'Sparks'.

Mike Robertson, Radio Officer

CHAPTER ONE

THE EARLY YEARS AND THE 1930S

As far as can be reliably established, the first mention of a permanent system of permanent coastal signal stations appears to have been made by the joint board of land and sea officer in 1785, under the presidency of the Duke of Richmond. Their report runs as follows;

"Your Majesty's Land and Sea officers beg leave to recommend a set of signals be established on the projecting headlands from Land's End to North Foreland with intelligent mariners to work them as an essential advantage in conveying early intelligence from one end of the Channel to the other of the approach of an enemy and for the protection of commerce".

This recommendation was introduced at the outbreak of the French Revolution and by 1795 thirty stations were in operation. Just before the Treaty of Amiens in 1802, this number had increased to 87, while between 1803 and 1814, during which the war had been continued, the number of stations had increased to 192.

There is no indication of how the 'intelligent mariners' were selected or what tests they had to undergo, but presumably they were Naval personnel who had been trained aboard H.M. Ships and who had been sufficiently intelligent to survive the rigours of the service.

The 1815 Act enabled the Admiralty to select and purchase sites for the erection of signal and telegraph stations, and while there is no evidence that these Admiralty stations were used for commercial purposes, Lloyds were in 1811 widely distributing signal logs and, as one of the objects of these stations was the 'protection of commerce', presumably the Admiralty stations were also used for signalling the movement of Merchant Ships.

Communications between each station and its Headquarters (either the local Naval base or Admiralty London) was by a series of hilltop semaphore stations and these were very efficient; for example, it is reported that 64 signals were sent from London to Dover in ten minutes.

After the war with France, most of these stations were abandoned, only those at the principal Naval ports being maintained. Due probably to the increasing importance of the shipping trade, private commercial stations were set up. At first these were at the approaches to the Thames (already established) and the Mersey (station established at South Stack in 1827).

In 1852 the Electric Telegraph Company set up a signal station at Lymington to report the arrival of mail steamers to Southampton, this station was eventually transferred to Hurst Castle.

Again no information is available as to the tests required for a post as an operator at such a station, but presumably the employees of the Electric Telegraph Company must have been capable of sending Morse code at fairly reasonable speeds.

The Electric Telegraph Company was taken over by the Post Office in 1870, including the operation of their coast signal stations. The department found the operation of these stations to be uneconomical and lost money on the deal. An agreement with Lloyds was entered into to operate all the department's signal stations, and in 1888 the Lloyds Signal Station Act became law, allowing Lloyds to acquire sites for further signal stations. In 1891 they signed an agreement with the Admiralty whereby Lloyds took over the working of the stations under Admiralty control, on condition that they could send commercial traffic over the system. The Admiralty had the right to take over all stations in an emergency but agreed not to establish any signal station on the coast for commercial signalling. Lloyds appeared to consider this agreement to be binding not only on the Admiralty but also other Government departments, and that they had been given a monopoly of commercial signalling. A letter from the Post Office to the Board of Trade in 1895 stated that Lloyds had for practical purposes the monopoly of signal stations in the country. Most of the Lloyds stations were operated by ex-Naval personnel and telegraphists but as far as can be ascertained there were no examinations required.

March 1896 saw a young Italian by the name of Marconi present a letter of introduction to Sir William Preece, the engineer-in-chief of the Post Office. This letter gave brief details of a system of 'signalling without wires', which gave rise to numerous experiments during the next few years. Preece had previously experimented with the 'conduction' principle of signalling, using the conductive properties of sea water over distances of up to a few miles, and was therefore most interested in this 'new' phenomenon.

The first experiments with the Post Office took place in 1897, with test signals being made on Salisbury Plain and between Lavernock (South Wales) and the island of Flatholm in the Bristol Channel. This was followed by an operational system working between Bournemouth and the Isle of Wight, a distance of about 12 miles. In 1898, wireless equipment was installed on board the East Goodwin lightship off Ramsgate and the South Foreland Lighthouse, and these tests became the first link between a ship at sea and a station on land.

By 1903, the Marconi Company had equipped several stations around the coast of the United Kingdom for the exchange of telegrams with ships, and ship owners made arrangements with Marconi for shipboard equipment to be installed and for trained operators to be carried as crew members. Many of these 'pioneer' radio operators were Post Office telegraphists, who were attracted by the new service. The Marconi coast stations were located at Browhead (Crookhaven), Rosslare, Withernsea, Caister, North Foreland, Holyhead, St. Catharines, The Lizard, Malin Head, Inishtrahull and Liverpool.

The first British ship to be fitted with Wireless Telegraphy was the S.S. *'Lake Champlain'*, and in those days the Marconi engineer who fitted the equipment to a ship usually sailed with her as the operator. This system of supplying operators for ships continued until 1903 when the Marconi Company set up their first depot and school at Seaforth Sands. Future Radio Officers spent approximately three months at this school acquiring a knowledge of the equipment and a Morse code speed of about 20 words per minute; The Marconi Company had no set standard of examination for their future employees and the officer-in-charge of the school wrote a few technical questions on the blackboard and gave a practical Morse code test. If he was satisfied with the standard, the student was then appointed to a ship and began his career as a Radio Officer.

The S.S. 'Lake Champlain' – the first British vessel to be equipped with Wireless Telegraphy.

Part of the Lloyds/Marconi agreement was that Lloyds undertook not to communicate with ships using apparatus other than Marconi equipment. This produced an incident that went some way to bringing about the first International Conference of Wireless Telegraphy. Prince Henry of Prussia crossed the Atlantic from America on the *'Deutschland'* (fitted with German equipment), and on approaching the United Kingdom he tried to send a Wireless message to the German Emperor via one of the Lloyds/Marconi stations. The message was refused.

This action naturally offended the German Emperor and although the companies apologised, the incident drew the attention of the German Government to the fact that a situation dangerous to their interests might arise if Wireless Telegraphy stations were allowed to discriminate in this manner. They accordingly invited various administrations to join in a conference on the subject and this produced in 1903 the first International Conference on Wireless Telegraphy, held in Berlin. Representatives from Germany, Austria, Spain, USA, France, Great Britain, Hungary, Italy and Russia were in attendance for this historic first meeting.

An act of Parliament in 1904 saw the Postmaster-General become the licensing authority for all stations equipped with Wireless Telegraphy equipment (both shore-based and on a British ship). This act in effect started the Post Office's involvement in Wireless Telegraphy services, and also made it illegal for any person to install or work Wireless Telegraphy apparatus in the United Kingdom or on British ships in territorial waters without this licence. This act was extended in 1908 to cover British ships on the high seas.

The commencement of the Post Office Wireless Telegraphy service can be traced back as far as July 1908, when Capt. F.G. Loring (RN) was appointed Inspector of Wireless Telegraphy. Capt. Loring had been Officer-in-Charge of the Naval Coast Wireless Stations since 1902, and his appointment came about as a result of the first International Convention of 1906, which spurred the government of the time into developing a ship-to-shore Wireless Telegraphy service.

An interesting extract from a letter written by a Mr. David Cook of the Marconi Company to Sir William Preece of the GPO from 1907 shows the future of Wireless Telegraphy communications in their eyes, and even casts very early doubt on the future of Radio Officers;

"In the existing state of the art, an expert telegraph operator must be carried on every ship. This will, however, be changed within a period of time measurable in months. Wireless telephony working with apparatus of great simplicity will take the place of the existing plant; word of mouth will supersede the Morse alphabet; the skilled operator will no longer be required: a ship's officer after undergoing a very short training will be able to adjust the apparatus and send a receive messages with much greater ease than flag and lamp signals can now be used. The lower cost of telephonic installation will lead to its adoption on the meanest tramp steamer and in sailing ships. We are, in fact, quite close to the time when wireless communications will supersede flag and other systems of visual signalling in the Mercantile Marine of the world".

Remember – this was written 5 years before the '*Titanic*' disaster, and some 92 years before the implementation of GMDSS.

The Post Office built a station at Bolt Head, near Start Point in Devon, which was opened in November 1908. Prior to this date, the Marconi Company and Lloyds had operated stations at Crookhaven (near Bantry Bay, Ireland), The Lizard, Niton (Isle of Wight), Caister (Norfolk), Seaforth (Liverpool), and Rosslare (South-East Ireland). Lloyds' owned stations at Malin Head (Northern Ireland) and North Foreland (Kent), the former being operated by the Marconi Company on Lloyds' behalf. The original invitation for applicants to man these stations was opened to seagoing operators, laying down age limits of 21 to 24. These stations become the forerunners of today's coast radio stations.

Artist's impression of the Bolt Head Wireless Station/GBA.

Equipment at these stations was similar to that installed at the Lizard Radio Station, which was:

"A hut containing the apparatus which generally comprises a 1-1/4 H.P. oil engine belt driving a 0.9 kW D.C. dynamo charging accumulators, energising a 10-inch coil. The transmitting apparatus is arranged for the production of a 300 metre wave by means of the usual Marconi jigger, and Leyden jar condenser. The receiving apparatus consists of the Marconi magnetic multiple tuner and recording receiver"

In June 1909, the Post Office made proposals to the Marconi Company and to Lloyds to buy the coast stations, and after negotiation, this became the property of the Post Office on the 29th of September of that year. Twenty Marconi operators transferred to Post Office employment and were supplemented by the recruitment of telegraphists from within the Post Office. Staffing at each station consisted of an Officer-in-Charge and four operators. The total number of ships then fitted with Wireless Telegraphy was 619, of which 286 were British. The existing tariff of 10½d per word was maintained, although in 1911, reduced rates were introduced of 5d per word for ships voyaging not more than 200 miles from the UK, and 5d per word for ships voyaging between 200 and 1000 miles from the UK, or to ports in the Baltic.

Apart from the control of coast radio stations, the Inspector of Wireless Telegraphy was now responsible for the examination of operators, the observance of the International Radio Regulations, and the inspection of licensed stations.

The main purpose of the coast stations was, of course, to assist with the safety of life at sea, and the first recorded instance of this occurred on 18th April 1910, when the S.S. *'Minnehaha'* ran aground of the Scilly Isles. The official Postmaster-General report quotes;

"At 12.52 a.m. on Monday the 18th instant, the Lizard wireless station heard the 'SOS' (distress) signal from the *'Minnehaha'* a large steamship of the Atlantic Transport Company, which was aground on a rock in the Scillies. The instructions issued by the Inspector of Wireless Telegraphy for the guidance of operators of this kind were carried out, and the Lizard station was instrumental in obtaining help for the distressed ship from Falmouth. Communication between the ship and the coast station was maintained throughout the morning and was resumed on the two following days. All the passengers were landed safely on one of the smaller Islands of the Scilly group, though this was apparently effected by means of the local lifeboats.

This is the first occasion since the acquisition of the Marconi coast stations by the Post Office on which a call for assistance has been addressed to a coast station direct by a ship in danger. We have every reason to be satisfied with the result of this first test of the usefulness of the Post Office stations in the case of accidents to shipping".

This was to be first of literally thousands of distress cases handled by the Post Office coast stations over the next 90 years or so, and it is impossible to estimate the number of lives saved through the involvement of these stations, both short-range and long range.

Traffic levels increased steadily as more and more ships became fitted with Wireless Telegraphy equipment; by 1912, 1,948 ships carried wireless apparatus, of which 778 were British. Further recruitment from within the Post Office became necessary to man the coast radio stations, and 200 applications were received. Twenty-four selected volunteers were to go to Hunstanton and Skegness (two Naval Stations) for two months training in the 'rudiments of the art' and eighteen of these were to be taken on immediately.

Meanwhile, apparatus was being renewed and brought up to date. Cullercoats Radio was opened in March 1912, and in September 1913 the station at Rosslare was transferred to Fishguard. Tariff structures were changing on a regular basis, and in 1914 the 3d per word rate was applied to all trawlers, irrespective of the length of their voyages.

During the Great War, the coast radio stations became under the control of the Admiralty, but continued to be operated by Post Office staff, who became part of the R.N.V.R. (Shore Wireless Service). Details of operations during the war are difficult to find, although it is known that the development of new apparatus (such as the thermionic valve) occurred during these years. At the end of the war, Admiralty stations at Wick, Portpatrick, Grimsby and the Lizard D.F. were taken over and modernised, later shutting down Grimsby – which had taken the place of Caister – and opening a new station at Mablethorpe. Lizard D.F. station was closed in 1931.

The Post Office was also responsible for the inspection of ship's wireless equipment, and one interesting letter (dated 2nd November 1918) sent both to the Marconi Company and Siemens Brothers highlighted one problem;

"I am directed by the Postmaster-General to inform you that it has recently been necessary to take disciplinary action in several cases where wireless operators have been found asleep while on duty, thus imperilling the safety of the ships on which they were serving. He finds it necessary to issue a general warning that in any further case of the kind which may be brought to his notice the Certificate of the operator concerned will be suspended. I am to ask that you will be so good as to bring the matter to the notice of the operators in your Company's employment".

It would appear that the excuse advanced by a number of Operators in cases of which the above letter was the outcome, was the lack of proper sleeping accommodation on board the vessels on which they were serving.

With the resumption of normal working, preparations were made for the reorganisation of the service. Telegram tariffs were increased slightly to 11d per word (Full Rate), 5½d per word (Intermediate Rate) and 3½d per word (Low rate). By 1920 there were 3,754 British ships fitted with Wireless Telegraphy out of a world total of 12,622. Staff at each of the coast stations was increased from four to eight operators. Wick Radio, a station built by the Admiralty during the war, was taken over by the Post Office early in 1920, and the stations at Bolt Head and Crookhaven were closed, the latter being replaced by Valentia Radio, another station built during the war.

To cater for the overwhelming demand for ship-to-shore communication services, an old army complex in Devizes was taken over by the Post Office in 1920. The location of the site was at Morgan's Hill, and the history of the complex is interesting:

A Marconi Imperial Wireless chain receiving station was built on the south-east slopes of the hill in 1913, as the receiving station for the Leafield transmission station in Oxfordshire. The contract for the Imperial Chain was terminated by the Post Office after the outbreak of war and the station was not used for its original function when Leafield was resurrected after the War.

The structure was converted to military use in 1916 for the Royal Engineers as an army intelligence station, used to determine the position of German Zeppelins and communication stations. The plans for the structure showed masts 300 feet (91m) high and 2,700 yards (2,500m) long. It is not known when the station closed but sources show that it was considered redundant by 1919.

The USA Federal Communications Commission (FCC) 'Radio Service Bulletin' dated January 2nd 1920 formally announced the arrival of the station with a brief paragraph;

"The British Post Office advises that the Government station at Devizes, call signal GKU, is open for communication with ships by means of continuous wave Wireless Telegraphy. The wavelength used is 2100 metres. The station will normally be open at all times for the receipt of messages from ships, but it will, as a rule, call ships only during the following periods: 1 to 1.30 a.m., 5 to 5.30 a.m., 9 to 9.30 a.m., 1 to 1.30 p.m., 5 to 5.30 p.m., and 9 to 9.30 p.m., Greenwich Mean Time. The station will not, however, be available for communication during the period of 3 minutes every half hour commencing at 15 and 45 minutes past the hour (GMT) as British ships using wavelengths other than 600 metres for any special purpose are required during these periods to keep watch on a wavelength of 600 metres. The rates of charge for messages sent through Devizes Radio are the same as those for messages through other Government shore stations of Great Britain."

However, questions in Parliament were raised over the manning of the new service as reported in Hansard on 29th April 1920:

Major MORGAN asked the Postmaster-General whether, in accordance with his undertaking, all the Post Office wireless stations in England, Wales, and Scotland were returned to the Post Office about two months ago; whether, during the period since the Armistice, the Navy and instituted commercial working at purely Naval stations such as Grimsby, Dover, and Rame Head, at which commercial traffic is being handled by Naval ratings, although the Post Office wireless staff and wireless stations are capable of handling the work; whether the long distance ship and shore service, which was recently inaugurated, and which is worked from a Naval station near Aberdeen, falls into the same category, seeing that the work is of a purely commercial character; and whether, in view of the importance that commercial traffic should not be handed over to the Naval Department, he will make representations to the Parliamentary Secretary of the Admiralty with a view to having the whole of the commercial wireless traffic dealt with by civilian operators at Post Office stations?

Mr. ILLINGWORTH: The Post Office stations in England, Wales and Scotland were returned about three months ago. The Post Office has also taken over the Naval station at Wick and will take over the Naval stations at Grimsby, Dover, and Port Patrick, The Naval station at Rame Head is available for commercial work as a relief to the Post Office station at Land's End, but little or no commercial work is actually done there. A certain amount of long distance commercial work with ships has recently been carried on at the Naval station at Aberdeen, because no suitable Post Office station is available; but a station is being erected near Devizes, and when this is ready—as I hope it will be in a few weeks—the Aberdeen station will no longer carry on commercial work.

Major MORGAN asked the Parliamentary Secretary to the Admiralty whether he is aware that in April, 1919, the Parliamentary Secretary to the Admiralty gave an assurance that there was no intention that the Admiralty should permanently retain any station in which Post Office wireless operators were employed; whether, as a result of that I assurance, all the Post Office stations have been returned to the control of the civil authorities; whether during the period since the Armistice the Navy has established commercial working at purely Naval stations such as Grimsby, Dover, and Rame Head, and that such commercial working is being handled by Naval ratings despite the ability of the Post Office stations and staff to cope with this traffic; whether the long distance ship and shore service recently inaugurated is being worked from a Naval station near Aberdeen although this traffic is also of a purely commercial character; whether there is a strong element at the Admiralty who favour the permanent retention of commercial wireless working by that Department; whether the Post Office inspector of Wireless Telegraphy and his immediate subordinate are ex-Naval officers; and whether, having regard to the necessity for the removal of all purely commercial traffic from the control of the Navy, he will cause immediate inquiry to be made as to the reasons which have led to the retention of certain commercial services under the control of the Naval authorities?

Sir J. CRAIG supplied the following particulars:

Stations in Home Waters. All General Post Office stations have been handed back to the General Post Office, with the exception of two stations in Ireland. Naval stations in the vicinity of Naval ports accept commercial traffic in addition to Naval traffic in cases where the General Post Office considers that the amount of traffic would not justify the erection of special stations for commercial work. The Naval Station at Aberdeen is being used for long distance commercial work at the request of the General Post Office, as they have no station as yet capable of dealing with this traffic until the General Post Office station at Devizes is completed.

Stations Abroad. Naval stations abroad are only used for commercial traffic at the request of the Governments concerned. The Admiralty are averse to undertaking any commercial work and are only doing so to assist communication when no other suitable stations exist, or in the event of cable breakdown. The last part of the question does not therefore arise. The Post Office inspector of Wireless Telegraphy and his immediate subordinate are believed to be ex-Naval officers."

Shortly before the opening of the station, the local press reported:

"A long distance wireless station is shortly to be opened at Devizes. Tests are taking place (says the Wireless World) and the coming into commission of the new station will mark an improvement in wireless communication from shore to ships at sea. This station was already partly constructed for the old Imperial scheme and was used as an intercepting station during the war. The station has now been converted into a transmitting station, it has been erected under contract with the Post Office, and is constructed for the purpose of communication with ships at sea. Messages have been received by the Marconi Company informing them that good readable signals have been obtained on ocean liners at distances of 1,600 miles. The aerial is supported on two masts, 300 feet high."

Local historian Arthur Cleverly recalls;

"The site of Devizes Radio (or Devizes Wireless Station, as it was always known), was located at Morgan's Hill, just to the north of the town. The site is steeped in history, most notably being adjacent to the location of the battle of Roundhay Down, which took place during the English Civil War - the last Royalist Victory.

There had been an army presence on the site from the 1860s onwards, and it was used as a rifle range prior to the Post Office purchase in 1920. Old lead rifle bullets can still be found in the area if one looks hard enough!

Several huts were dotted around the area, belonging to the site maintenance staff. These were small, brick-built affairs, with iron pipe chimneys, and equipped with an iron stove, known as 'tortoise' stoves. After the site was closed, these huts were let to 'weekenders' who camped on the hill and used the huts for cooking and storage. Unfortunately, none of these huts have survived.

There is little left that remains of Devizes Radio apart from a length of heavy steel chain, one end set in concrete, which would be the 'ground' end of a steel guy line to one of the masts".

The increased range of the new service encouraged passengers and crew of the transatlantic liners to send radiotelegrams when several days from port; similarly, vessels trading around the ports of Europe could use the service on a regular basis.

Initially staffed by 10 radio operators, traffic levels grew considerably as more and more ships became equipped with suitable apparatus; this necessitated the installation of additional equipment and the separation of transmitters and receivers to provide duplex (transmitting and receiving on different frequencies) channels.

An article published in the February 1924 edition of 'Experimental Wireless', extracts from which explain the operation of the station in great detail:

"Devizes is the Post Office station which deals with long distance ship traffic. The ordinary stations are equipped with 1.5 to 5 kW spark transmitters operating on 600 metres, and are capable of working ships at ranges of from 150-300 miles, depending on conditions.

In August 1920, however, a service was inaugurated, working on 2,100 metres c.w., for the purpose of maintaining communication with the large Transatlantic and southbound liners, at considerably greater distances. The chief ships operating on these long distance routes were accordingly fitted with 1.5 kW valve transmitters and the necessary c.w. receiving gear, while Devizes, which was chosen for the shore station, was equipped with a 6 kW valve set. Since that time, the number of ships fitted with c.w. apparatus has increased enormously, and there are several 3 kW valve and 5 kW arc sets in operation.

Devizes Radio in 1920.

The Devizes station is situated in the middle of the Wiltshire Downs, about four miles from Devizes itself, on the Marlborough Road. Two masts only are in use at the present time. These two masts are each 300 ft. tubular steel masts of the Marconi pattern and are 600 ft. apart

The aerial is a double-cage, 250 ft. long, each cage consisting of three wires spaced equally round a hoop 3 ft. in diameter. The lead-in is a 6-wire cage, 4 ft. 6 ins. diameter. The earth system consists of 56 plates, 6 ft. by 2 ft. 6 ins. by 24 s.w.c. copper, buried in the earth on the circumference of a circle 50 ft. in diameter.

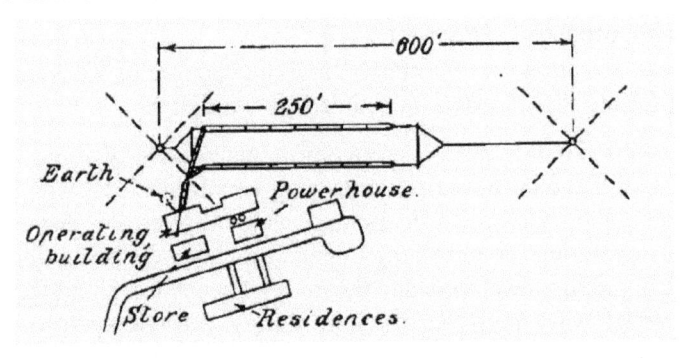

Devizes Radio Station operational layout.

The transmitter is a Marconi 6 kW valve set, employing six M.T.4 oscillating valves and four M.R.4 oscillating valves. Power is supplied to the set at 500 volts 300 cycles, and is stepped up to 15,000 volts before being applied to the rectifying valves, which thus supply H.T. to the oscillating valves at about 10,000 volts D.C.

The receiving apparatus consists of a Bellini-Tosi radio-goniometer, the signals from which are passed through a two-stage HF filter, amplified and rectified on a Marconi type 55 amplifier, and finally passed, if conditions permit, through a two-stage note filter.

Since its inception, the traffic handled by the station has increased enormously, and it is proposed to install a second transmitter at Devizes and to remove the reception point elsewhere, so enabling duplex working to be carried out with both transmitters by remote control from the receiving station".

Another station at Rugby (Warwickshire) was established at this time; this station was provided with higher power, higher masts, and used longer wavelengths (16 kHz, call sign GBR). This station provided one-way (broadcast) communication to ships. Publicity material then available gave details of costs;

- Rugby Radio to ships on all seas at 1/6d per word.
- Devizes Radio and other land stations for ships up to five days from a British port, per word; facilities also available from ship to shore.

In the early 1920s, the Post Office became interested in point-to-point communication. Stonehaven Radio (near Aberdeen), using Poulsen arcs, conducted services with European countries from 1920 until 1926. In 1922, a service between Leafield (Oxfordshire) and Abu Zabal (Cairo, Egypt) was introduced. These two stations had been built as the first links in the Empire chain of wireless stations to provide radio communication throughout the British Empire, a chain that was never completed due to the outbreak of war in 1939 and the improved technology which made the initial design obsolescent. However, this 'point-to-point' service (or 'beam' service, as it became known) was to prove a vital starting point for the development of a global ship-to-shore service on the higher frequencies, as links between the U.K. and Canada, South Africa, India, and Australia were conducted on frequencies between 9 and 19 MHz, using highly-directional aerial arrays.

One point-to-point transmitting station was located at Dorchester, with an associated receiving station at Somerton, some 30 miles away. Both stations would in the future become involved with the Portishead Radio complex, but at the time of opening (16th December 1927), the Dorchester station provided a route to New York followed shortly by a route to South America. Services to Japan and Egypt were provided by the end of 1928.

The aerials at this site were of extreme interest and importance; these 'beam' aerials provided highly-directional transmitting paths, ideal for point-to-point use. An array of stacked dipoles produced the 'beam', suspended from triatics attached to 90 feet wide cross arms on top of 287 feet high lattice steel masts. These masts were spaced at 640 feet intervals, with the width of the beam aerials being some 500 feet. The actual length of the dipoles depended on the wavelength in use, the range covered being between 15 to 75 metres (4-20 MHz).

These aerials were taken down in 1966, being replaced by Log Periodic aerials for the Associated Press service for the Eastern, Middle East and South African routes, the latter closing in 1975. All point-to-point services were closed or transferred from Dorchester by 1970, although the site provided Maritime Radio services for Portishead Radio until July 1979, when the site closed for good.

Although the 'point-to-point' service is outside of the scope of this book, references will be made to it from time to time when circumstances become relevant to ship-to-shore communications involving Portishead. This 'point-to-point' service was resurrected to some degree in the 1980s as part of Portishead Radio's Gateway service.

Meanwhile, coverage of the United Kingdom's coast improved with the opening in December 1921 of Portpatrick Radio to cater for ships off the west coast of Scotland.

A glimpse into the operating procedures at Devizes was published in the FCC 'Radio Service Bulletin' from 1922;

"The difficulties which have been experienced at the Devizes station in the reception of long distance radiograms from ships are found to be increased by the fact that the wavelength used by ships frequently differs appreciably from the wavelength nominally used for communication with Devizes, viz, 2,100 metres c.w. Recent observations showed that the wavelength actually used varied from 2,080 to 2,180 metres.

The use of different wavelengths by the ships not only necessitates a continual adjustment of the receiving apparatus at Devizes, but also increases the probabilities of interference, and may even lead to a call from a ship being missed if the station is at the time listening to another ship which is using a different wavelength.

It is accordingly proposed to arrange for Devizes to send out every four hours a long dash lasting a minute on a standard wave of 2,100 metres c.w., to enable the ships to check, and adjust if necessary, the wavelength of the transmitter on board. It is essential that ships equipped with c.w. apparatus be accurate in their adjustments. In order to avoid encroaching on the time available for the exchange of traffic, it is proposed to arrange for the standard wave to be emitted toward the end of the period of 10 minutes, viz, 35 to 46 minutes past the hour GMT, during which ships using long wave c.w. are at present required to keep watch on 2,400 metres. Devizes will as at present broadcast on 2,400 metres the call signals of the ships for which messages are on hand, and they will ask ships to "stand by" until 44 minutes past the hour for the standard wave of 2,100 metres. This standard wave will be emitted at 0044, 0444, 0844, 1244, 1644, and 2044 GMT."

A novel way of communicating with ships was devised by Cunard in 1922, which was reported on 13th September 1922 as:

"A novel wireless link across the Atlantic was reported by the Cunarder *'Mauretania'* which arrived at Southampton yesterday. Whilst mid-ocean she received and relayed messages to the *'Aquitania'*, then on the point of entering New York Harbour. These messages, which were despatched from the wireless station at Devizes, were received and passed to the American coast by a chain of Cunarders, no fewer than six which were in touch that particular day."

Very much an early version of QSP (relay of messages free of charge) traffic.

The Devizes station became one of the pioneers in maritime radio communication, and in 1924 the first-ever 'automatic' ship-to-shore message was handled, as reported in the press on 29th February:

Devizes Radio Station c. 1925.

"During the voyage of the White Star liner *'Olympic'*, which arrived at New York from Southampton on Wednesday, a trial was made for the first time on shipboard of transmitting wireless messages automatically from sea to London, so that these could be actually printed in the General Post Office without human aid. The experiment was successfully achieved at 90 words to the minute by a high-speed transmitting apparatus, at a distance of 700 miles from the land station at Devizes, the printing of the messages at London being perfect."

Correspondence from ex-staff member Bill Welch has provided the names of the complete operating staff at Devizes at around 1925 (other staff would have been employed at the station but left before that time, details of whom have not been located):

- Ernest F. 'Chippy' Greenland (Officer-in-Charge)
- Frank P.J. Dockrill (Radio Officer – Second-in-Command)
- Arthur W. Hicks (Radio Officer – ex Land's End)
- George Steer (Radio Officer)
- Percy Crisell (Radio Officer)
- George Winchcombe (Radio Officer)
- Robert Blacklock (Radio Officer)
- H Gibbs (Labourer/Engineman – transferred to Portishead)
- J Adams (Labourer/Engineman – transferred to Portishead)
- H A Dixon (Labourer/Engineman – transferred to Portishead)
- Mark Little (Labourer/Engineman)
- Mr. Gillett (Labourer/Engineman)

Percy Crisell, who moved to the Burnham site after Devizes closed, was apparently an excellent operator who could send Morse code impeccably at 30 words per minute and even faster after a pint or two!

From 1925, the following staff were employed:

- Charles Addis
- Richard L. Ryan (transferred to Portishead)
- George McManus (transferred to Portishead)
- Charles S. Ingram
- Harold Lear (transferred to Portishead)

- Gwilym James (transferred to Portishead)
- William (Bill) Welch (transferred to Portishead)
- H. H. Wadsworth
- H. Masters (joined the BBC)

When Mr. Greenland left, the Officer-in-Charge became George Woodward (ex-Portpatrick Radio) who eventually became 2nd-in-charge at Portishead. He was succeeded by William Smallwood and then Francis Crank.

Many regular ships that communicated with Devizes were the *'Majestic/GFWV'*, *'Olympic/GLSQ'*, *'Berengaria/GBZW'* and *'Carinthia/GLKY'* amongst others.

The growth of Devizes Radio became so rapid, that on January 29th 1925 the Post Office opened a new purpose-built receiving station in Highbridge (near Burnham-on-Sea, in Somerset) to cater for the demand.

Later that year, new transmitters were constructed at Portishead near Bristol, a move which brought about the now internationally-famous name of 'Portishead Radio'. The site was carefully chosen to ensure non-interference between the transmitting and receiving sites, and it also had the advantage of being located on the Burnham-Bristol cable route. Keying of the transmitters was effected over landline connections from the receiving site at Highbridge.

Three transmitters and three receivers were installed at the new sites, still operating in the range 110-160 kHz, which permitted several ships to be dealt with simultaneously.

The earliest photograph of the Highbridge Receiving Station building, c. 1925.

The new building at Highbridge was originally a single-storey affair, with the operating room being located on the left-hand side of the entrance passage which had a sliding door. When the first floor was added later in the decade, work would not have interfered with the normal operation of the station.

An article by F. Hollingshurst in the December 1925 issue of 'Modern Wireless' describes the new station succinctly:

"The building itself is a single-storied brick structure and contains a power room, battery room, two store rooms, instrument room, landline room, office for the officer-in-charge, and retiring room for the staff. The instrument room is at the back of the building and overlooks the open site on three sides, thus allowing free access for leads from the aerials. Between the instrument room and the landline room is a communication hatch through which messages received and written by the wireless operator can be passed for retransmission by another operator over landlines to London or through which messages received from London can be passed to the wireless operator for transmission via Devizes to ships.

The receivers and other apparatus controlled by the operators are placed on benches which run round three sides of the instrument room, and are so arranged that the tuning, direction finding and filament controls of the receiver, the switch by which the transmitter at Devizes is started up and shut down, and the hand key which operates the transmitter are all within easy reach of the operator's chair.

Highbridge Radio Officers at work c. 1925.

A few feet away is the telegraph sounder by which landline signals from the engineer at Devizes are made audible. Power boards are fitted at intervals round the wall, and from these, current at 52 volts for radio high tension circuits, 26 volts and 28 volts for landline circuits and 4 volts for radio low tension circuits is taken through fuses. These boards are also fitted with terminal connections to the radio and telegraph earths. A low tension switchboard for connecting the three low tension batteries for charge or discharge is also fitted in this room.

The receiving equipment originally supplied by the Marconi Company at Devizes is in use at Burnham. Directional reception is employed to minimise interference from the East, the majority of the reception being from West and South-West. Facilities are provided whereby traffic may be received from ships at high speed and relayed directly to the Central Telegraph Office in London, where it is recorded telegraphically.

The transmitters at Devizes are remotely controlled from the receiving point at Burnham. This remote control apparatus at the receiving station consists essentially of a hand key and a switch key. The former is used for sending Morse signals to the engineer or for transmission through the sets to ships, and is arranged to reverse the connections between the battery and the line. The switch key is used for starting and stopping the transmitting sets, and in the 'stop' position connects a 26 volt battery through the hand key to the line, while in the 'start' position it changes over to a 78 volt battery".

Staff of Highbridge Radio Station. c. 1926.

Back Row (L-R): George Steer, 'Taffy' Davies, William Williamson, J. Brazier, Wallace Winchcombe, Percival Crisell, Jack Lovibond.

Front Row (L-R): Arthur Spence, Cyril P. Arundel, Unknown Smith, William MacGregor, Charles W. Dawson.

1926 saw the first high frequency ship-to-shore communication service from the new sites at Portishead and Highbridge, with world-wide tests being carried out with the cruising vessels '*Carinthia*' and '*Olympic*'. These proved to be very successful, and these tests showed that the range of radio communication could be practically world-wide, but this range could be affected by radio propagation conditions, the time of day, and the location of the vessel. It was also clear that short-wave transmitters did not need the high power required for their long wave counterparts, and that efficient transmitting and receiving aerials could be erected in a much smaller area.

Much of the work on investigating the use of short-wave communication had been carried out by radio amateurs, whose experiments had shown that great distances could be achieved with simple equipment and basic aerials. Post Office engineers continued to develop suitable receiving and transmitting equipment which was used at their various radio sites across the UK.

Three of the six masts at Devizes, which had become a significant landmark in the area, were taken down and reported in a somewhat poetic fashion;

Highbridge Receiving Station with new first floor completed c. 1926.

"In the early part of September, three crashed to the ground, sacrificed on the Altar of Progress. It seems particularly fitting that this site should have been chosen for the scene of destruction; the offspring of the most modern branch of Engineering laid low amidst the ruins of the most ancient! Local antiquaries will tell you, if you are interested, of the great feats of those Neolithic Engineers, who, at Avebury, ages before men learnt to use iron implements, built those huge circles of stone within whose confines sun-worshippers conducted their religious rites. They will speak with enthusiastic voice of the mighty deeds of the civil engineers of ancient times, who, to restrain the predatory encroachments of hostile Lords, excavated the great dyke that skirts the boundary of the Radio Station site. These deep ruts crossing the line of masts, mark the road made by Roman military engineers, and many an invading chariot has passed this way in the great days of old.

The trails of the engineer have left their mark unobliterated even by the passing of Time! Many other tales will be told, but if you pass down from the hill into the villages below the story you will hear most often concerns the sad passing of those 300-foot masts".

One of the Devizes Radio masts after felling.

The article continues;

"Picture the scene as the anxious moment arrived to put the scheme to test! Holding a sledgehammer aloft, the executioner stands above the pin. A word of command and the heavy tool swings quickly down. The pin flies out, the stays slacken and fall in towards the mast. For a moment the 'stately and air-heaving tower' remains erect, and then a great shout goes up. "She's falling! She's toppling over!" Slowly at first, but with ever-increasing speed, the structure heels about the base.

And now, the story of the passing of the masts is told. If, on that mid-September night when the last mast fell, you had chanced to pass by the great stones in the quiet village of Avebury, you would have seen – as the moon rose – small dark figures flitting about the ground, and heard a whispering sound rising and falling in gentle cadence. You and I, cold matter-of-fact engineers, would have dismissed the matter with some crude remark about shadows, and wind blowing leaves about. But the Poet, with wiser discernment, will tell you that these were no shadows caused by the rustling leaves, but elves and goblins jauntily capering around the great hoary stones, chanting joyously their triumphant song."

Trees and steel will bend and bow,

Wood and clay will wash away,

Build it up with stone so strong;

Huzza! 'twill last for ages long.

Some recollections of the station are fascinating to read. Former employee Bill Welch remembers:

"At Devizes, the call sign was GKU in both long waves (2013 and 2100 metres – we didn't think in terms of frequency in those days). At Portishead, we had a couple of extra-long wavelengths, 2479 metres and one other, all using call sign GKU.

The first short-wave transmitter at Devizes in 1929 had the call sign GKT (36.54 metres) plus another around 18 metres. The first ship we worked was I think the *'Jervis Bay'* (Hudson and Commonwealth Line), call sign VZDK – the first ship to be experimentally fitted with short-wave. It was a great thrill at the time that we could work her while she was berthed in Perth harbour. On one occasion we worked the *'Jervis Bay'* in Perth when her signals faded out. When the signals returned the R/O on the vessel advised "don't worry, it is only our donkey engines working and reducing the power from our generators!'

We had originally six Marconi tubular masts (315 feet) arranged in line, erected after the 1914-1918 war for the proposed Imperial Chain (I think the link would have been to Abu Zabal (Egypt). Around 1926 three were felled as they were no longer needed.

In 1927 the night staff were marooned for a day or so by snow. Shift changes were 0830, 1630 and 2300."

The GPO also utilised the existing high power transmitters at Rugby for the service, as described in an article from 1926:

"The Post Office has just inaugurated a service for sending wireless messages to ships at sea all over the world. This, it is claimed, makes most important advances in the uses to which Wireless Telegraphy may, in the future, be put. The Post Office has for a number of years, had shore stations all round the coast, which transmitted messages to ships to a range of about 300 miles. With the development of Wireless Telegraphy, it was found possible, however, to establish a station at Devizes for sending and receiving messages up to 1,500 miles, and at a later date, the Leafield station was started with a range of 3,000 miles. But even that had been found insufficient, and through the big power station at Rugby it is now possible to send messages to ships all over the world. In order to achieve this, the transmission is centralised in the Central Telegraph Office in London. Messages are sent midnight and noon, and the operator provided with a loudspeaker, by means of which he is assured that the emission from Rugby is satisfactory. Owing to the great power which has been employed in the transmission messages over such distances, and the impossibility generating such power on board ship, it is not possible to get an acknowledgement of the receipt of the messages. The officials in charge of the service, however, are hoping to attain 95 per cent proficiency in the new venture. All the private messages sent out at noon are repeated at midnight, and those transmitted at midnight are sent out again at noon the following day, so as to obviate as far as possible the prospect of the messages failing to reach their destination through atmospheric or local disturbances.

The first messages on this new service were sent to a ship at Port Arthur just after midnight on Saturday. That this message was received satisfactorily the officials in charge are confident, since it has been reported from the Dutch East Indies that the signals from Rugby are being received perfectly. Perforated tape is used for the transmission, in order to ensure absolutely perfect formation of each unit to be put into the ether, transmission being fixed at a standard at which any reasonably trained operator could receive. The wavelength which messages are being sent is 18.700 metres. There is to be no fundamental change in charges for private messages as a result of the new system. From the coast stations, the charges range from 11d to 1s 1d a word, from Devizes 1s 2d a word, and from Rugby 1s 6d a word."

On a lighter note, in October 1926 the newspapers reported an attempt by a certain individual to send a message to the planet Mars;

"It was learned in London during the evening that from the Government high power station at Rugby a message had been arranged to be transmitted to Mars just before midnight. This did not mean that the Government the Post Office authorities were inclined to regard communication with the planet within the bounds of possibility. The message had been handed to the Central Radio Office of the GPO and had been accepted for transmission, reception not being guaranteed. It was of a length which would take somewhere about three to four minutes to transmit by Morse, and was described as being of 'no known language'. The transmission was to take place on a wavelength of 18.248 metres which the Rugby station happened to be using last night, and the message was to be prefixed with three M's. In the Morse code, two dashes represent the letter M. At five minutes to twelve the Rugby station transmitted into the void the message handed in by a private person. An official at the Central Radio Office discussing the matter with a Press representative said the message would probably be charged for at the long-distance ship rate. viz. 1s 6d per word. "The Rugby station has a world-wide range," added the official, "but I do not know about the planetary system. If people wished to send out messages and are prepared to pay for them there does not seem to be any valid reason why the Post Office should refuse revenue". Asked if the authorities would agree to transmit a message intended for the moon or the man therein, the official repeated that he could see no reason for rejecting such a message."

It is not known if the message was received.

On the 13th October 1927, the station was involved in a rescue operation that made international news.

"Ruth Elder, the first woman pilot to set out from New York to Paris by air, ended her flight less than 1,000 miles short of her destination with her plane in flames in the open sea. But the woman flyer and her co-pilot Captain George Haldeman, were taken on board the Dutch tanker *'Berendrecht'*, safe and sound.

A message from the Dutch steamship, relayed by a British steamer to the Devizes Radio Station, said that the crew of the plane were saved with both well but the plane was destroyed by fire. The position of the *'Berendrecht'* was given as 43.34 north, 21.39 west, about 325 miles north northeast of the Azores and it was expected that the daring flyers would be landed there."

It was becoming clear that the Devizes station was quickly becoming unsuitable for the amount of traffic being handled, coupled with the unreliability of the equipment being used. Hansard reported during a Parliamentary session on 11th April 1928:

Sir H. BRITTAIN asked the Postmaster-General whether he is aware of the great inconvenience which is caused to passengers on transatlantic vessels owing to the unsatisfactory service from the Post Office wireless station at Devizes; that owing to the frequent difficulty in securing communication with this station a very considerable proportion of messages sent to this country from ships in the Atlantic have to be transmitted from the vessel to America, and thence forwarded direct to this country either by cable or wireless, involving considerable increase in the cost of the message; that vessels like the *'Majestic'*, which are provided with high-speed apparatus for transmission and reception, find it almost impossible to work with Devizes, which is only capable of working at hand speed; and what steps he proposes to tale in the matter?

Sir W. JOYNSON-HICKS: I am aware that there is some difficulty in receiving messages at the Devizes wireless station from the Western Atlantic owing to interference. To meet these difficulties a new receiving station is being provided in the neighbourhood of Weston-super-Mare, and on its completion I anticipate that these difficulties will disappear. At the same time, high-speed apparatus is being installed.

Sir H. BRITTAIN: Can the hon. Gentleman say when this apparatus is likely to be completed?

Sir W. JOYNSON-HICKS: I cannot say at the moment. We are pressing on with it.

It was clear from the above that plans for the new transmitter site at Portishead were being actively pursued.

History was made when the first commercial wireless message ever despatched from an aeroplane flying over the Atlantic was handled by Devizes, and reported in the national press on 30th June 1928:

"Captain Courtney completed the first stage of his flight to New York on Thursday, when he flew to the Azores from Lisbon. He started at 8 a.m., and landed in Horta at 6 p.m., thus taking 10 hours for the flight of about 900 miles.

Courtney is using a Dornier Wal flying boat. He carries wireless, and sent out messages during the flight. One message was: "Five hours out from Lisbon. Both engines running well. Crew very cheerful. All going well. Courtney, en route Azores, 1 p.m.".

At 2.30 p.m. Mrs. Courtney received from her husband the first commercial wireless message ever despatched from an aeroplane flying over the Atlantic. The message was picked up by the Devizes wireless station and transmitted in the ordinary way by the Post Office. The message was as follows: "Five hours out. Love to you and Olive. Frank".

Olive is the daughter of Captain and Mrs. Courtney.

From Horta, Courtney proposes to fly to Newfoundland or Halifax, Nova Scotia, a distance of 1,800 miles, and then to New York. If he succeeds he hopes to make a direct flight back from New York to England. Should Courtney reach New York, his flying boat will the first man to fly from Europe to there according to plan.

Courtney, whose parents were Irish - from Cork - has been devoting his time to the development of the seaplane for some years past. He believes that the transatlantic air services of the future will be by seaplane."

As a result of tests conducted from the new transmitter site at Portishead, the ship-to-shore radio service began to develop along two separate lines;

1. The short-range service, including distress, from the normal coast radio stations, and;
2. The long range world-wide service conducted from Burnham/ Portishead.

Portishead transmitting station masts and aerials.

The latter service commenced in earnest in December 1927 on 119 kHz on high power and on 143/149 kHz with 6 low power transmitters. The transmitting aerials were supported on 4 x 300 ft lattice steel masts of 8 ft square cross-section, and were carefully designed to minimise interaction at such close frequencies. As a result of these new transmitters becoming operational, the Devizes transmitting station was finally closed in 1928, although tests on the short-wave bands continued until 1929. The 'Radio Services Bulletin' dated July 31st, 1928 reported:

"Devizes station has been discontinued, and the services formerly carried on by that station are now being operated by the Portishead station, call signal GKU, wavelength 2,013, 2,100, and 2,479 metres, c. w., location about mile southeastward of Blacknore Point Light, in longitude 2° 47' 30" W., latitude 51° 28' 41" N"."

Use of the 'short-wave' band (wavelengths below 100 metres) began to be increasingly used on board ships. In 1928 the number of ships fitted to work this particular type of transmission and reception was about half a dozen. This number grew rapidly and by 1930 the Post Office Guide gave particulars of 222 ships fitted to communicate on short-wave.

In 1929, Burnham receiving station maintained a watch on short-wave for a few hours a day with a single receiver and a borrowed transmitter. Despite the increasing number of ships being equipped with short-wave equipment, the service grew very slowly until 1933, when the introduction of a cheap letter telegram for ships saw traffic figures increase yearly.

The short-range service continued to expand, and between 1932 and 1935 all the coast radio stations were re-equipped with valve transmitters following the Washington radiotelegraph convention of 1927 which abolished spark transmitter operation. In general, each coast station was equipped with a main and emergency transmitter and a minimum of two receivers. The daylight range was approximately 300 miles, whilst a much greater range was possible during hours of darkness. Transmitting and receiving aerials were erected on the same site, and only simplex (transmitting and receiving on the same frequency) working was possible. Caister Radio closed down and was replaced by a temporary station in Grimsby, and subsequently by a new station at Mablethorpe (to be known as Humber Radio/GKZ). Fishguard Radio closed in 1934 and its services transferred to Burnham/GRL.

It is intriguing how the call sign GRL became assigned to Burnham. Originally, GRL was assigned to Rosslare (RL = RossLare) but when the station transferred to Fishguard and subsequently Burnham, the call sign remained.

Radiotelephone facilities were provided in 1928 to cater for the growing number of small craft fitted with suitable apparatus, and on 1st March 1934 Seaforth Radio opened the first radiotelephone link between ships at sea and telephone subscribers ashore. In these early days, the coast station radio operator had to manually switch between the ship and shore side to allow each party to speak alternately, a situation which was remedied by duplex channelling in later years.

The Post Office was keen to develop their maritime radio services, and in 1929 a well-publicised event took place, which was widely reported in the press at the time:

"When Mr. A.S. Cauty, general manager of the White Star Line, was a day out on the journey to New York by the *'Homeric'*, which left Southampton on Wednesday last week, it became necessary for an urgent message to be sent to him from the head office of the White Star Line, London, which demanded an immediate reply. The Post Office, in conjunction with the Marconi Company, was able to arrange a combination of landline telephony and Wireless Telegraphy whereby the desired discussion was possible. The Post Office provided a special telephone line connecting the White Star Office, London, to the Burnham Wireless Operating Station. A Post Office operator stood by at the Burnham end of the telephone line to write down the message passing from London to Mr. Cauty, and to receive his reply. Beside this operator sat another listening in to the transmissions from the *'Homeric'*. This second operator also controlled a key, which operated the high power transmitting set situated at Portishead, some 40 miles away, which formed the wireless telegraph link between Burnham and the *'Homeric'*. This is the first time that Wireless Telegraphy and landline telephony have been linked in this manner to provide quick two-way communication between a person ashore and a person on board ship at sea. The arrangements worked very smoothly and satisfactorily. The separation of the operating station at Burnham and the high power transmission station at Portishead is quite interesting. An arrangement is developed by the Post Office to enable the operators at these busy stations to communicate simultaneously, and with an entire absence of interference with one another, with ships scattered over the Western Ocean route, the South American route, the South African route, and the Mediterranean and Far East route."

The early 1930s saw Portishead Radio handle traffic from the large 'flying boats', with their onboard Radio Officers sending radiotelegraphy traffic from all over the world, most notably from South Africa and the Indian sub-continent.

Details of the station at the time appeared in various radio magazines, which also gave some fascinating technical details:

"The long range ship service is taken care of from Portishead (transmitting) and Burnham (receiving) stations. In addition to the long-wave installations, there are two short-wave transmitters that can communicate with ships in any part of the world. These short-wave transmitters operate on 8,210 kc/s (36.54 metres) and 16,840 kc/s (17.81 metres). The tube used in these transmitters consists of a 354 kW silica valve with a tuned circuit in series with the grid and anode. This special oscillating circuit was developed by the Post Office for this service and has the merit of great simplicity, and together with other features included, gives a frequency constancy well within the requirements specified. The supply current, rectified and partially smoothed out, gives a characteristic 600 cycles note. The 16,840 kc/s transmitter is coupled to a horizontal array with a 2-wavelength aperture mounted on a lattice steel girder which is capable of rotation on a vertical axis. The rotation of this beam system and the keying of the transmitters are controlled from the receiving station at Burnham. For the 8,210 kc/s transmitter a half-wavelength vertical dipole is used as a radiator.

At the Burnham receiving station in addition to the long-wave receivers there are short-wave receivers for the 36 and 17-metre ranges. As a matter of comparative interest, it should be mentioned that the long-wave receivers are provided with large frame aerials and verticals providing a cardioid diagram for directional reception, and have a six-stage high-frequency amplifier, separate oscillator, and also low-frequency band filters. A rotating array similar to that at Portishead is provided for the directional reception of the 17-metre service. The use of a beam antenna for reception and transmission gives a substantial improvement on any short-wave service, and the use of a rotatable system has solved the problem of adapting the directional array for ship service. After locating the ship desired by means of the receiving system, the transmitting array is directed on the ship by remote control from the receiving station. Horizontal dipoles with open transmission lines are used for the 36-metre service."

Don Mulholland remembers the rotating beam well:

"As a young lad, I took my Dad's lunch box to the station. He (Robert Mulholland) was one of the first six operators who started in 1925. In the station, on the first floor, one of the operators lifted me up on a chair by a window - he put my hand on a control and told me to turn it. The contraption in front of me started to rotate. It was rectangular and mounted on wheels with a pole on each corner. From the poles was suspended a directional dipole with reflector. As it was rotated, a similar contraption at Portishead also turned. In those days only liners were fitted with short-wave and they collected/distributed traffic with ships in their area on 500 kHz. The liners were worked on schedules and the aerials would be suitably rotated before the schedule started. The system worked only on 36 metres (8 MHz)."

Don continues:

"Another thing I remember from those days was the aerial system. It was fed to the station on poles and each aerial entered the station through an arrangement mounted in the walls. It was not very satisfactory.

My Dad made a suggestion that the aerials should enter the station through holes cut in the windows; he got an award for the suggestion".

An FCC 'Radio Service Bulletin' from 1932 issued an interesting procedure with regard to the traffic lists from Portishead:

"In order to facilitate the reception of call signs, Portishead Radio now emits the call signs of the following ships in alphabetical order: *'Europa'*, *'Bremen'*, *'Ile de France'*, *'Paris'*, *'Berengaria'*, *'Homeric'*, *'Majestic'*, *'Mauretania'*, *'Aquitania'*, *'Olympic'*, *'Empress of Britain'*, *'Leviathan'*; followed by the call signs of other ships for which traffic is on hand, also in alphabetical order;

Ships' operators on British ships are invited to report from experience, through their employers, on the suitability of this arrangement which is designated to meet the convenience of ships.

Ships' operators are reminded that the earlier clearance of traffic in both directions may often be effected with consequent reduction in the length of traffic lists, if the opportunity is taken to call Portishead Radio some time before scheduled periods of traffic lists. The non-utilisation of periods of light loading, e.g. before the 1030, 1230, and 2230 lists, frequently leads to congestion following these times.

Ships' operators should not rely primarily upon traffic list periods to effect clearance of messages, but should endeavour to effect communication with Portishead Radio, if their position suggests practicability of such contact, as soon as traffic is handed in on board; or whenever, irrespective of traffic list times, it is thought that traffic for them may be on hand at Portishead Radio."

It is not known whether this was an experimental procedure, or one which became normal practice at the time.

Portishead Transmitter Building, 1928.

In 1934, after maintaining the same tariff for 16 years, the 'intermediate' and 'low' rates were increased to 6d and 4d per word respectively. In addition, a Ship Letter Telegram (SLT) service was introduced, in the ship-shore direction only, at a charge of 3d per word with a minimum of 5/- for twenty words.

Meanwhile, maritime use of the high-frequency Wireless Telegraphy service continued to grow, and technical facilities at Highbridge and Portishead were improved. The original HF transmitter, operating on 18/36 metres (GKT/GKC) was supplemented by 4 20kW transmitters of Post Office design in 1936. In addition, primarily for the cruise liner *'Queen Mary'*, a rotating beam aerial was constructed; four medium-sized telegraph poles mounted on a lattice platform supported the aerial, driven around by a sprocket and chain mechanism. As you will have by now observed, the tariff structure for telegrams was becoming increasingly complex, and a rationalisation took place in 1936. The 'full rate' per word charge was reduced to 8d per word (except for messages exchanged with certain foreign ships), and the 'intermediate' and 'low' rates were merged into a single rate of 4d per word applicable to all ships on regular voyages not exceeding 1000 miles from the UK.

With the exception of certain parts of the Pacific, the short-wave service covered practically the whole of the world's oceans beyond a radius of about 2000 miles from the station. Within this radius, and serving chiefly the North Atlantic and parts of the Mediterranean, the medium distance service provided the necessary channels.

Rotating beam aerial for communication with the Atlantic liners.

Before a message could be passed to a ship, direct contact naturally had to be established. To obtain this contact, Portishead Radio transmitted, at 2-hourly intervals, a list of ships' call signs for which messages were waiting. A ship hearing its call sign on this list could then contact Portishead to receive its traffic. Ships wishing to send messages, however, could of course contact Portishead at any time. Messages received at Portishead were forwarded to the London Central Telegraph Office (CTO) for onward delivery to the destination.

RUGBY RADIO for distant Ships on all Seas **1/6** per word

TELEGRAMS by **DAY & NIGHT TO SHIPS**

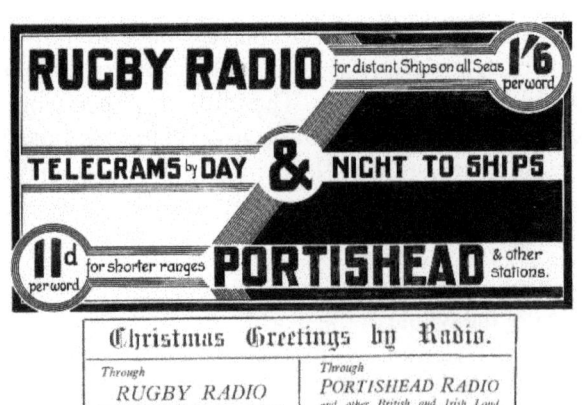

11d per word for shorter ranges **PORTISHEAD** & other stations.

Christmas Greetings by Radio.

| Through
RUGBY RADIO
TO SHIPS ON ALL SEAS
at 1/6 per word. | Through
PORTISHEAD RADIO
and other British and Irish Land Stations to Ships up to 5 days from any British Port,
at 11d. per word. |

Facilities are also available from Ships to Shore.
For particulars enquire at any Post Office or Ship's Radio Office
Hand in your Christmas and New Year Greetings early.
Special arrangements have been made for all messages passing through British Stations, and consisting clearly of Christmas and New Year Greetings only, to be delivered as far as possible on Christmas Day and New Year's Day respectively.

BRITISH POST OFFICE WIRELESS SERVICES.

Some Post Office advertisements of the 1930s.

However, despite the efficiency of the new station at Portishead, not all locals were happy with the service. A letter dated June 14, 1928 to the Postmaster-General written by Albert V. Brice, the proprietor of the Royal Hotel at Portishead, stated:

"The incessant interference caused presumably by the Portishead Wireless Station on the 1600 metre wavelength, is now, if anything, worse than ever. On my set it practically obliterated the Prince of Wales' speech yesterday morning, and it spoils the programme every day from the 5XX* station, so that it is quite useless taking the trouble to tune in.

I am now taking up the matter with the Postmaster-General, as it is intolerable that one should be forced to pay a yearly licence for wireless reception, and then this interference should be allowed to completely spoil the programmes. I am informed that an apparatus can be installed in the interfering station to stop the trouble, and the least that can be done is for that to be done without delay, and put an end to an intolerable nuisance. I shall be interested to hear what other steps are being taken in the matter."

*5XX was the Daventry long wave transmitter of the BBC.

However, a student at the South Wales and West of England Radio Training College, Len Adlard, remembers this era with affection;

"We used to look on Portishead as our own special station on which we practised our Morse receiving. The call sign was then GKU, but the operator always sent it as 'GTAU', lengthening the first dash of the 'K'. Our landlady would not let us put up an aerial, so we used to hook up our crystal receivers onto the wire mattresses of our beds with excellent results".

Another Radio Officer of the time, V.J. Hickey, remarks in his book 'Time to go, Sparky';

"A day out of Pernambuco (Brazil), and keeping the usual schedule with Portishead Radio, I was delighted to receive a message from my parents, wishing me a happy eighteenth birthday. I told Portishead "That one's for me" and asked him to hold on a moment for my reply. Portishead wished me many happy returns and said he would phone my reply through personally to my home town of Minehead. My father had cycled the mile into Minehead and called at the main Post Office to send my birthday telegram. After that, he went to the bank and rode home. He was just getting off his bike when the telegraph boy hailed him. You can imagine his utter astonishment at receiving a reply to the telegram he had sent only twenty minutes before!"

As now, propagation conditions often made communication difficult from certain parts of the world, and it was not uncommon for ships to call for long periods before contact with Portishead was established. However, it was quite usual to see adjacent receivers at Highbridge handling traffic with a ship trading in Pacific waters and a passenger liner leaving New York, which confirmed the world-wide range of high-frequency radio.

Some high-frequency communication was directly concerned with the saving of life and property, and an example of such an episode was related in the December 1948 issue of the Post Office Telecommunications Journal;

"An interesting case was that of the vessel which ran aground some hundreds of miles up the River Amazon. The ship was out of range of the local stations in the area and could not make contact with the shore. The distress message was passed directly to Portishead on high frequency and then forwarded by cable to the local sea rescue organisation".

High-frequency radiotelephony services were available to ships at this time; however, it was the Post Office receiving stations at Baldock and Bearley who handled this traffic, utilising transmitters at Rugby (Warwickshire). Point-to-point radiotelephony services had been in operation on the high frequencies since June 1928 from the same sites, so the maritime service was a logical progression. Use of this new high-frequency R/T service was confined mainly to the large passenger liners, although it soon became commonplace for merchant vessels to use this facility.

By 1930, a second, lower power (60 kW) long wave radiotelegraphy transmitter had been installed, using call sign GBV and operating on 78 kHz. It utilised a 'T' shaped aerial supported between two 820 ft masts. One use of this transmitter was to broadcast weather charts and forecasts to the R100 Airship when flying between the UK and Canada.

That same year, a short-wave radiotelephone service commenced to passenger liners in the North Atlantic; tests with the RMS *'Olympic'* started in January, and results were extremely positive. This resulted in a full commercial service becoming operational from February, transmitting on 4.975, 8.375, 12.780 and 17.080 MHz, with carrier power from the transmitter at Rugby 'A' Building being between 3 and 4.5 kW. The call sign used was G2AA, and early users of the service included the liners *'Olympic'*, *'Majestic'*, *'Homeric'* and *'Leviathan'*. The return signal was received at the Post Office Receiving Station at Baldock.

A merchant seaman with Elders and Fyffes between 1931 and 1934, E.M.Shaw, recalls how Portishead Radio helped supplement his income when his ships docked at Avonmouth;

"I well remember being told by some of the able seamen that they would like a voyage off to paint the masts at Portishead. They were paid 1/- for every 100 feet, so once the top of each 300' mast was reached they were paid 3/-. At that time an apprentice was paid 1/- a day at sea, whilst a seaman got £8.12/- a month".

Early postcard of the Portishead transmitting site.

The mid-1930s saw the station become increasingly busy on all frequencies, which resulted in the station becoming well-known both in the maritime world and within the Post Office. In order to provide some well-deserved publicity for the service, the station manager at the time (E.F. Greenland) wrote an excellent article for the April 1935 issue of the 'Marine Observer', extracts of which are quoted below;

"The wireless operators at Burnham, as at other British coast stations, are skilled telegraphists drawn from the staff of the Post Office Telegraph Services. It is this personnel who make immediate contact with the seagoing operator, the two constituting the link through which the service is provided. On land, however, as at sea, skill in telegraphy is only the first step towards full operating efficiency and the gap between the two covers a long period of operating experience in all its phases. One of the most important lessons to be learned is that of appreciating the other fellow's difficulties. At sea, the 'Old Man' may be hard to please and require 101 things to be done while the shore operator is becoming caustic at the lack of attention, having an important message on hand. On shore there may be several ships waiting at a busy period, when everyone is hard at it and trying to give the more distant man a chance in case he fades out and can't clear his message, while the ship operator threatens to offer his traffic through a foreign station. A good example however of the basic co-operation and good fellowship which exists in the mobile service is the patience and forbearance with which both sides treat the 'young hand' at the game".

The article then goes on to describe the equipment in use at Burnham;

"The medium wave service is handled by three large receivers covering the 2,000-2,500 metre wave bands, and a receiver for the 600-800 metre range. These are provided with facilities for all-round reception or for reception only in a certain sector if required to cut out jamming and interference. This directional reception in a certain sector cuts out also, of course, signals in other sectors. Another receiver is available for the small craft coastal telephone service, which is provided to enable small craft not carrying wireless telegraph operators to exchange telegrams with the shore. Out of a total of 14 receivers on the station, the remaining 9 are installed for the short-wave service which has grown so rapidly in the last few years that it now carries more than half the load of the station. These are associated with a dozen or more short-wave receiving aerials directive and otherwise. The directive types present a somewhat novel feature in mobile working as previously their use was confined to long-distance services between fixed points. Judicious arrangements of the directions in which they face, and the angles or sectors covered, have permitted their successful employment to serve the main shipping routes of the world. By their use, at both transmitting and receiving stations, it has been possible to compensate to some extent for the power and equipment limitations that are automatically imposed on ships' sets. Their effectiveness is found in the regular and usually daily communication which short-wave ships in the most distant parts of the world".

A similar article from the 'Wireless World' published at the same time, goes into further detail:

"At Portishead, there are three medium-wave and two short-wave transmitters. The former work up to about 2,000 miles, the latter to world-wide ranges.

The medium-wave transmitters have an input power of 15, 6 and 6 kilowatts respectively, the short-wave transmitters have each 10 kilowatts. The medium waves are transmitted from T aerials supported by 300 ft masts, the short waves from dipoles, or directional arrays (of which there are 10), or a rotating beam aerial (whose rotation is worked from Burnham), according to the positions of the ships communicating and the waves in use. The medium waves used are in the 2,000-2,400 metre band, and the short waves in the 18, 24 and 26 metre bands. In addition, there is equipment for sending on the ordinary coast station band, and there is a telephone transmitter for small craft communications.

GPO

RADIOTELEGRAMS TO H.M. SHIPS.
A NEW SERVICE
via PORTISHEAD RADIO.

On and from 1st December, 1935, radiotelegrams will be accepted at any Postal Telegraph Office for

H.M. SHIPS ON FOREIGN STATIONS
(EXCEPT THE MEDITERRANEAN STATION)

at an inclusive charge of

7ᴅ· A WORD.

Messages should be routed through PORTISHEADRADIO.

RATES
for
RADIOTELEGRAMS TO H.M. SHIPS
as from 1st December, 1935.

Location of ship	Appropriate coast station	Inclusive charge per word
HOME WATERS	CLEETHORPESRADIO	3d.
MEDITERRANEAN	*RINELLARADIO	6d.
	⊕GIBRALTARRADIO	9d.
ALL OTHER STATIONS	PORTISHEADRADIO	7d.

* For ships in the vicinity of Malta and in Eastern Mediterranean.
⊕ For ships in the vicinity of Gibraltar.

Publicity for the new radiotelegram service rates, 1935.

The receiving equipment at Burnham consists of three medium-wave and nine short-wave receivers, and the coast station and telephone receiver. The aerials for the medium wave receivers consist of Bellini-Tosi loops; those for the short-wave receivers being similar to the short-wave transmitting aerials, including a rotating beam aerial.

Short-wave working with ships is still much more erratic than medium wave working, and though its range may be considered as world-wide under favourable conditions there are certain areas, such as the North and South Pacific, where difficulty is nearly always experienced.

The transmitting station at Rugby is the largest Post Office station, and is, indeed, the largest station in the world. Messages for ships in any part of the world are broadcast by telegraphy, the actual keying being carried out in the General Post Office in London. Official Press messages are broadcast thrice daily, news agency messages four times daily, and private messages for individual ships twice daily. These messages are sent on an 18.750 metre wave on Rugby's main transmitter of 350-kilowatt input, and are simultaneously sent on a short-wave transmitter from the station at Leafield (Oxford) so as to give ships all over the world the best possible chance of reception.

A weather shipping statement is broadcast from Rugby twice a day and a time signal is sent out at 6 p.m. These transmissions are made on long wave only.

Rugby Radio, it will be noticed, is used only for broadcasting messages to ships, as far as telegraphy is concerned, as does not deal with any telegraph traffic from ships. It's most important use is, of course, for point-to-point communications, not communications with ships.

Rugby is also used for a telephony service with certain Atlantic liners. The subscriber on shore is connected, through the radio terminal at the Faraday House Exchange in London, to Rugby Radio when speaking and to Baldock Radio when receiving, and conversation is carried on in the ordinary way over the telephone so far as the subscriber is concerned. Communication with the ship is first established by telegraphy through Portishead Radio, and arrangements made for putting through the call at a certain time. Short waves are used for this telephone service, five waves being utilised between 17 and 94 metres, according to the time of day, the season of the year, and the range. This service can be extended through London to many countries all over the world; and a passenger in mid-Atlantic has no difficulty in holding a conversation with, say, a friend in Sydney."

On 1st December 1935, the GPO announced a 'New Service via Portishead Radio' which was to provide radiotelegrams to H.M. Ships. This was publicised by a poster distributed to telegraph offices and Post Offices throughout the UK:

Operating staff at Portishead Radio were recruited from experienced and qualified marine radio operators, and this continued until the outbreak of war in 1939. Over 3.5 million words of traffic were handled during the year, a figure which had increased rapidly since the start of the decade. However, with a conflict imminent, safeguards to protect the service had to be installed. To ensure continuity of service, power to the transmitters at Portishead supplied by the North Somerset Electric Company was augmented by a standby diesel generator installed underground beneath the tennis court.

Operating area, Highbridge, mid-1930s.

If further evidence was necessary as to the amount of traffic handled at the station, a 1936 newspaper report provides such:

"Although the *'Queen Mary'* failed at her first attempt to win the blue riband for the Atlantic crossing she has made one record in the amount of telegraphic work she gave the Post Office. Some 48,000 words, mostly Press messages, were handled through the Portishead wireless station. This is the highest total sent from any ship, and it was over half of the total traffic last week at the Portishead station, which again established a record. In addition, the Post Office handled 54 broadcasts, for the BBC, America, France, Holland, and Denmark. Those for America were relayed through the transatlantic telephone service, and for the Continent by submarine cable. If these figures are not reached on the return voyage of the *'Queen Mary'*, it will probably be because it has not the same interest for America, but the telegraphic and broadcast traffic should again be heavy in view of the assumption that, under favourable conditions, the ship will successfully attack the record for the crossing."

Highbridge receiving station staff c. 1935.

Back Row (L-R): William Rogers, Pat Arundel, Bill Wallis, Joe Brazier, Harry Bateman, Fred Sherratt, Wally Blow, Charles (Jube) Dawson.

Middle Row (L-R): Herbert Barron, Mr. Davies, Eric Atkins, Joe Wynne, Reg Hawkins, Bill McGregor, Cyril Chapman, Samuel Clarke, Percy Crisell, Clifford Gillingham (messenger).

Front Row (L-R): Thomas Norrish, Patrick Kenney, Tim Williamson, Edward Rogers, Ernest 'Chippy' Greenland, H.B. Smith, Robert Mulholland, George Steer, Vic Allen, Don Williamson.

On 24th April 1936, the GPO announced a reduction in the rate for radiotelegrams. A speech by Major the Rt. Hon, G. C. Tryon, Postmaster-General, stated;

"The public knows too little of the ship-shore wireless services and the men who work them both on shore and on board ship. Occasionally we read some ocean epic in which the famous signal 'SOS' plays a prominent part. There is also the urgency signal 'XXX' indicating that a vessel is disabled or in difficulties, though not in immediate danger. Emergencies requiring the use of one or other of these signals numbered 178 in 1935, about one on every second day.

The Wireless Services played an important part in the recent rescue of Ellsworth and Kenyon, the airmen engaged in the American Antarctic flight. It was decided to instruct the *'Discovery II'* in the Antarctic at the time, to call at Melbourne, take aircraft on board, and proceed to the Bay of Whales. These directions were passed to the *'Discovery II'* by wireless through the Portishead Radio Station, and contact was maintained daily for the exchange of reports and instructions. Although wireless conditions became more difficult as the ship neared the Bay of Whales, a few hundred miles only from the South Pole, communication was maintained satisfactorily on most days. Ships in the Antarctic were invited by Portishead to listen at scheduled times for any wireless transmissions sent out by the missing airmen. The *'Discovery II'* left Melbourne on 21st December and effected the rescue at the Bay of Whales on 15th January.

Medical service is available to ships that do not carry a doctor. In the event of serious illness or accident on board, the Master can obtain advice from the shore through coast stations. A special international code overcomes any language difficulty.

Navigation and meteorological warnings issued by the Admiralty and Air Ministry respectively and broadcast through Post Office coast stations amounted last year to 180,000 words.

There are twelve Post Office coast stations in the United Kingdom and Irish Free State through which these and other services are given. In addition, the Post Office stations at Rugby and Baldock conduct special service with ships at sea.

All these stations are increasingly used for transmission and reception of private and business messages as well as for the special services to which I have referred. All of them can communicate by radiotelegraphy with ships within 300 miles of the shore. One of them – Portishead Radio – can operate services with a range that is world-wide.

The British Post Office handles, through its coast stations, more than twice as many messages as any other European country. Last year the traffic amounted to nearly five million words. Rugby Radio transmits private messages twice a day to ships which are out of range of the ordinary coast stations.

Small ships need carry only one operator, but big lines have sometimes a dozen. The *'Queen Mary'* has, I think, a wireless staff of fourteen. Every operator must have a certificate of proficiency from the Post Office examiners. Nearly 1,000 candidates presented themselves for examination last year, so the service is anything but unpopular.

The general usefulness and efficiency of these services make it a special pleasure for me to announce to you a reduction of rates which will be brought into effect on and from 1st July next. The new tariff, which is rather too detailed for me to give you fully here, and is indeed not yet quite ready in all its details, will be published a little later, but the chief item will be the reduction of the standard rate from 11d a word to 8d a word. This new rate covers messages to and from ships in all parts of the world.

Another important item for ships making distant voyages and not provided with wireless sets capable of transmitting over long distances is the reduction from 1s 6d a word to 1s a word in the Rugby rate. This Rugby service meets a small but steady demand for communicating with ships otherwise out of touch with the shore".

One amusing incident was highlighted by some internal correspondence within the Post Office from 1937; it appears that some of the tape wrapping around some of the aerial poles had sustained damage caused presumably by woodpeckers. The engineers were requested to renew the tape and treat the poles with tar, with all holes repaired accordingly. The Department of Agriculture and Horticulture at Bristol University were, however, quite excited by this and were keen to view the damage caused, as a letter from them to Mr. E.W. Hedges of the radio station states;

"Dr. Miles will be very pleased to see the poles near Highbridge that have been damaged by woodpeckers. He is away for the remainder of this week but has made a note of your telephone number and will give you a ring sometime on Saturday or Monday to arrange to see the poles with you".

It is not known if further damage took place after the above incident.

CHRISTMAS GREETINGS TO SHIPS BY RADIOTELEGRAPH

YOUR FRIENDS AT SEA

Send them your
GREETINGS
this Christmas by
RADIO

For ships reached through Portishead Radio or a British coast-station the charge is 11d. a word.
For ships reached through Rugby Radio the charge is 1s. 6d. a word.

HAND IN YOUR RADIOTELEGRAMS AT ANY POSTAL TELEGRAPH OFFICE.

Greetings may also be sent by radio from ships at sea to addresses on shore.

PG 204

Information card issued by the GPO in the 1930s.

On January 30th, 1937, the local Burnham-on-Sea newspaper carried a detailed article about the radio station, running to just over 2 pages. A lot of the article has little bearing on the activities of the station, but some relevant extracts are quoted below which are of useful historical interest;

"Referring to the ship-shore service in which the Burnham station participated, Mr. Read (Deputy Inspector of Wireless Telegraphy) said that it was not really appreciated what a large wireless activity there was in the Post Office. If the word 'wireless' was mentioned, everyone said 'Marconi' but actually the Post Office had a very wide telegraph point-to-point service. They had one of the biggest stations in the world at Rugby, and many telegrams went by the Post Office wireless service to the Continent, and not by Marconi. With the Rugby transmitters, they had a very heavy load and it was continually transmitting from about 10 o'clock in the morning to about 3 or 4 o'clock the following morning, all the year round. They were handling long official Press messages which went all over the Empire and the Colonies, and they also took a Press service for ships which covered the seven seas, besides a considerable press service for other parts. With all its services they had probably got an organisation which was as big as any other organisation in existence for wireless purposes. Coming back to the ship-shore services between the numbers of stations around the coast, he knew, and headquarters knew that it was second to none in the world. They had a very fine staff, and the operators were men who got on with the job. In Burnham and Portishead, they had what he thought was the finest long-distance station in the world. It handled the biggest traffic load of any ship-shore station in the world, and quite twice as much as any in Europe".

One of the original overseers, Mr. W. Williamson, was quoted as saying;

"It was exactly 12 years ago when GKU came into being, and they had seen a healthy child grown to recognition. There had been changes since then; one receiver had grown to 18, and there were nearly 40 operators instead of nine. The development of the station had only been possible by the kindness and co-operation of everyone at the station. The co-operation of the operators with him while had been overseer had been splendid, and it had been one perfect team spirit working all the time. He could have done nothing as overseer unless he could have relied upon everyone to do his best. The result was shown in the high regard which ships have for GKU. They had had some difficulties and stiles, but stiles were only made to get over. GKU was the premier of its kind as regards efficiency and service".

An internal document from 1937 gives fascinating technical details of the service at the time;

Portishead Radio Transmitting Station

Situated at Portishead near Bristol. All transmitters are remotely controlled from Burnham. Staff comprises one Engineer-in-charge and staff of 16 assistants.

Transmitters:

    1 medium-wave, input power 25 kW

    2 medium-wave, input power 6 kW

    (Wavelength 2013, 2100 and 2479 metres)

One of the 6 kW medium-wave transmitters is also equipped for working on 600 metres and 625 metres. Wave changing of this transmitter effected by remote control from Burnham.

1 low power telegraph/telephone transmitter for operating in the small ships waveband (100 – 200 metres)

3 short-wave crystal-controlled transmitters. Output power approximately 6 kW in each case. Equipped with quick-change aerial switches, operated locally by hand, for connecting to associated arrays and dipoles.

Array System:

Supported on 7 lattice steel masts 140 feet high. Directional on principal shipping routes. Wave bands covered 18, 24, 26 and 36 metres. One rotating beam aerial for 18 metres. Orientation of this aerial directly controlled from Burnham. Dipoles for omni-directional transmission available for all wave bands, and also 48 metres.

Operating Station

Situated at Burnham about 20 miles SSW from the transmitting station. Staff of 43 operators and 4 supervising officers.

Services – four distinct services are conducted.

1. A medium-wave telegraph service for ships within a distance of approximately 2000 miles (4 receivers)
2. A medium-wave telegraph service for ships within a distance of approximately 300 miles (2 receivers)
3. An intermediate wave telegraph and telephone service for small craft (coasters and trawlers) (2 receivers)
4. A short-wave service for ships in any part of the world. Eleven receivers are available for this service on various wavelengths according to requirements.

The aerials used are similar to those in use at Portishead for transmitting. All landline traffic is handled by means of teleprinters, a direct circuit to London being provided. Radio traffic handled during 1936 was as follows:

    Short waves   2,129,000 words

    Other waves   2, 102,000 words

The December 1937 Burnham Radio half-yearly report advised that a staff revision was introduced, with the short-wave facilities extended by increasing Search Points. The introduction of duplex working on 18 metres resulted in a greatly improved service being given to ships on this band.

Highbridge main building mid-1930s.

Of course, communication was not only limited to ships' business; in 1938 the Post Office experimented with position reporting of British Transatlantic Aircraft, a system that relied heavily on the watchkeeping of selected ships. In its basic form, both Portishead and Valentia Radio stations would maintain a day-to-day 'picture' of Atlantic shipping in the vicinity of the air route and, on receipt of notice of a projected flight, would obtain from five or six selected ships their accurate positions, course and speed and pass this information to Shannon Airport. This system was tried out during the first flight of the year and proved to be satisfactory.

About fifteen flights took place during the year, with additional flights being undertaken by French and American aircraft. Of course, other aircraft used the services provided by Portishead regularly, with some long-distance aircraft carrying dedicated radio operators for this purpose. The service continued during the war, but slowly declined as the Civil Aviation Authority (CAA) introduced new procedures and provided their own communication networks.

The Burnham Radio report from June 1938 indicated that there was a great deal of inter-ship working on the 36 metre band, causing considerable interference. Occasional inter-ship interference was also reported on the 24 metre band at times, but not as troublesome. Negligible interference noted on the 18 metre band.

Operating room with long-wave receivers on display.

Apparatus Room, Burnham receiving station, 1938.

The long-wave receiving equipment at Burnham. Morse key clearly visible.

Portishead transmitting site, c. 1938

An interesting atmospheric phenomenon took place in 1938 which was reported in 'Wireless World':

"At Burnham Radio, the station through which the short-wave ship-to-shore Portishead Radio is remotely controlled, the first indications of exceptional wireless conditions became apparent about 1750 GMT on January 25th. Abnormality was first marked on the 18-metre band, which is Burnham's main long-distance communication channel until 2000 GMT. A British liner approaching New York communicated with Burnham at about 1730 and commenced sending a batch of messages but by 1800 signals from this vessel had gradually faded to inaudibility, despite the fact that reception and transmission were being carried on with directional aerial arrays. As soon as this difficulty became apparent, Amagansett Radio, Burnham's 'opposite number' on the American seaboard was called, but without success. This Long Island radio station is usually so well received that immediate contact can be relied upon, but although Burnham ran an automatic calling slip for nearly two hours, no reply was forthcoming from either the liner or the USA station. Amagansett was just audible about 1900 GMT, but communications from the westward on its frequency continued to be nil, except that Palo Alto (California) Radio was readable on 18 metres at maximum strength, which was unusual. Communications on the 24-metre band were not noticeably adversely affected other than to the westward; the American coast stations were just audible. Conditions on 36 metres were variable, with intermittent periods of extreme fading and very heavy atmospherics; this state continued until about 0400 on January 26th. On the other hand, good communication was maintained on the 2,013-2,479 metre band. The spectacle at Burnham-on-Sea was remarkable. From west to north-east the sky to a great height was a dull pink in colour gradually changing to various tints of green."

View of the Highbridge receiving site in 1938.

The following December, the report stated that there was severe heavy atmospheric interference experienced on long wave. During the weeks ended 6th and 13th August atmospheric interference was exceptionally severe on long and short waves, and was considered to be the worst ever experienced on short waves. Seven dipoles were replaced by cage type dipoles and four additional cage type dipoles were fitted.

Operating console at Highbridge, December 1938.

The onset of war in 1939 brought considerable changes to the work of both Portishead Radio and the various other coast radio stations in the United Kingdom, and introduced new working practices which were maintained until the demise of the service some 60 years later.

CHAPTER TWO

THE WAR YEARS

The war years saw a great amount of activity and change at Portishead Radio. Commercial traffic ceased, as for a ship to transmit a message would invariably release its position to the enemy by direction-finding apparatus. Activity at Portishead and the coast stations was therefore confined to Admiralty traffic and the handling of distress calls and enemy position reports. These included news of the North Africa landings and the sinking of the Scharnhorst. Monitoring of the various clandestine stations from Europe was also undertaken, which ensured that the station was kept busy. In addition to the maritime services, a special aircraft section was constructed for communication with patrol aircraft in the North Atlantic. During 1941, arrangements were made for certain long range Naval services to be worked from Portishead Radio, and a Naval officer, with eighteen Naval telegraphists, were added to the staff of the operating station at Burnham. The Naval and Post Office telegraphists sat side by side, two to a point, each pair searching a particular wave band. If either heard a ship call, the operation was conducted by the Naval operator if it was from a warship, or by the Post Office operator if it was from a merchant ship.

The Admiralty expressed satisfaction with these arrangements and stated that they have resulted in a marked increase in efficiency in their communications with distant ships, especially warships.

A paper entitled Wartime Development and Services (written in 1946) gives an excellent insight into the role of both Portishead and the various UK Coast Stations during the conflict. It reads;

"During the war, considerable use was made of the Post Office Wireless Service and its personnel. Advantage was taken of all-round adaptability and the high degree of initiative which required in wireless operating. The following statement will give an indication of the varied tasks undertaken;

The opening of hostilities saw the cessation of ship-to-shore working for commercial purposes and 'silence-at-sea' was observed. The operators were required to maintain a vigilant watch during long periods of wireless inactivity; the operator had to be prepared to receive distress signals, signals concerning the approach of enemy aircraft, signals warning of the approach of enemy submarines, and any indications of illicit transmissions.

To meet difficulties in connection with submarine activity in the Atlantic, stations were opened in the Isle of Lewis, Isle of Tiree and Cape

Wrath. These stations were remotely situated and their success depended on the adaptability and efficiency of the staff. Engineering or operating assistance was not easily obtainable.

A station was opened at Stonehaven to cover the area off the East coast of Scotland.

When the all-out air offensive on the continent was begun it was assisted by the opening of a station at Ormesby for the major purpose of receiving signals from airmen brought down in the North Sea.

Accurate figures as to the number of airmen rescued as a result of distress signals received at Post Office Wireless Stations are not known, but it is thought that many hundreds were saved.

To assist in handling difficulties for ships entering Liverpool, a station was opened at Lytham St. Annes.

Post Office Personnel were used both at Burnham and at a temporary station to work alongside Navy personnel and maintain contact with both Naval and merchant vessels. Operational messages of the utmost priority were dealt with in a manner evoking warm commendation from the Naval authorities."

The December 1939 Burnham Radio report makes interesting reading. The conspicuous aerial masts and approaches to the station were repainted to camouflage them, and new huts were erected for army personnel on the site.

Prior to the outbreak of hostilities, considerable development in connection with transatlantic flights occurred. Flights became much more frequent and there was every indication, in view of the success of the flights and absence of casualties, that rapid expansion would take place. American aircraft on various routes were beginning to realise the usefulness of this station. They frequently established communication on short waves and cleared their traffic on several occasions.

The October 1941 report continued in a similar vein. Over 9 transatlantic flights were dealt with, and contacts with transatlantic aircraft flying between New York and Baltimore were established on several occasions. Indeed a record distance was obtained on 28th September at 2350 GMT when the Australian aircraft VHABF informed the station "departed Penang East bound 2340 GMT".

Shortly before the outbreak of war, a station was opened in Hertfordshire and was staffed entirely by Post Office Wireless personnel. This station subsequently formed the basis for four other stations employing a total staff of approximately 600.

The whole of the preliminary operating work in building up this highly secret organisation was undertaken by the Post Office, and was both highly skilled and requiring a considerable concentration and intelligence.

The received signals were both 'aural' and 'high speed'. In the case of the 'aural' signals they were in many different languages, as well as in code. They were transmitted on frequencies not intended for this country, and were often intentionally jammed by a broadcasting station. Despite these difficulties, and the inability to ask for repetitions, a very high percentage of messages was received. Perhaps it is of equal importance that a large number of 'operators remarks' were intercepted. The skill of the Post Office operator was demonstrated in his ability to read the particular characteristics in the types of Morse sending, and in his ready understanding of the language abbreviations used for 'telegraphist' conversations.

The 'high-speed' aspect of this work required an ability to tune in high-speed stations frequently transmitting in a direction to make reception in this country difficult. The messages were received on an undulator and had to be transcribed into type.

The effectiveness of this work, and the ability of the staff, were commented on by Mr. W.S. Morrison, Postmaster-General, and Mr. Alan Chapman, assistant PMG, and brought a personal commendation from Mr. Winston Churchill, Prime Minister.

To meet possible invasion, the United Kingdom was split into zones and linked by an Emergency Wireless Service. The initial training and opening of stations was undertaken by Post Office Wireless staff. In the event of bomb damage to a Coast Station, two mobile stations were held ready. The periodic testing was undertaken by Post Office Wireless Staff.

With the invasion of the Continent by the BLA (British Liberation Army), the Post Office Wireless Service handled press traffic from War Correspondents. In its early stages, this meant 'aural' reception of very weak signals transmitted from the Normandy beachhead. Later it involved the opening of a special station to handle thousands of words and after transcribing, retransmission on the teleprinter.

This vast development which finally brought in hundreds of loaned and temporary staff had, as its basis for efficiency and acceptance of responsibility, the one hundred established Post Office Wireless Telegraphists.

The necessity for a highly skilled and adaptable body of men was reflected not only in the work of development which we have outlined, but also in the demands for many of them to undertake work of a supervisory nature in the inspection of ships' apparatus".

These 'special service' stations were constructed and staffed for another Government department, employing some of the existing staff and recruiting inland telegraphists.

Another possible (but unconfirmed) use of the Highbridge site related to radio countermeasures as described below:

"During their operations against Britain, the Luftwaffe also made use of radio beacons located in France to assist aircraft navigation, and even against these radio countermeasures were employed. Instead of straightforward jamming, it was decided to receive and then re-radiate each individual beacon's signal from a transmitter located in Britain, thereby effectively falsifying the beacon's position. The transmitters involved were known as Masking Beacons or 'Meacons', the first having been installed in the region at Temple Combe in late August 1940, with an associated receiving station at Kington Magna. This was followed by a transmitting and receiving station opening at Highbridge in November, the transmitter from which was transferred to a separate site at Lympsham in March 1941.

The first success came some months later when, on the early morning of July 24th, the Lympsham transmitter, then re-radiating the Brest beacon, caused a Junkers Ju 88 of I/KG 30 which had been on a mission to Birkenhead Docks to land undamaged at RAF Broadfield Down, which at that time was still under construction. Low on fuel and thinking they were over France, the crew put down at 06.20 hrs, to make the first landing on what is now the main runway of Bristol's Lulsgate Airport, the Junkers subsequently entering service with the RAF as EE205."

So why was the radio station never attacked by enemy aircraft? The transmitting site at Portishead was indeed identified by the Luftwaffe in August 1940 as a possible target, with detailed information about the station noted on reconnaissance photographs. However, in common with Rugby Radio, it was likely that the transmissions acted as 'beacons' for navigation at night. To destroy such useful aids would be counter-productive. Similarly, messages transmitted by the station could be received and possibly decoded by the German intelligence, so why put this out of action? There is no evidence of any attacks on the receiving station, although some German bombers discharged their surplus payload over the Bristol Channel close to Burnham-on-Sea following attacks on Bristol. One bomb did fall on the Berrow Road in Burnham-on-Sea, destroying a house and killing the occupants, but it does appear that the town was not specifically targeted.

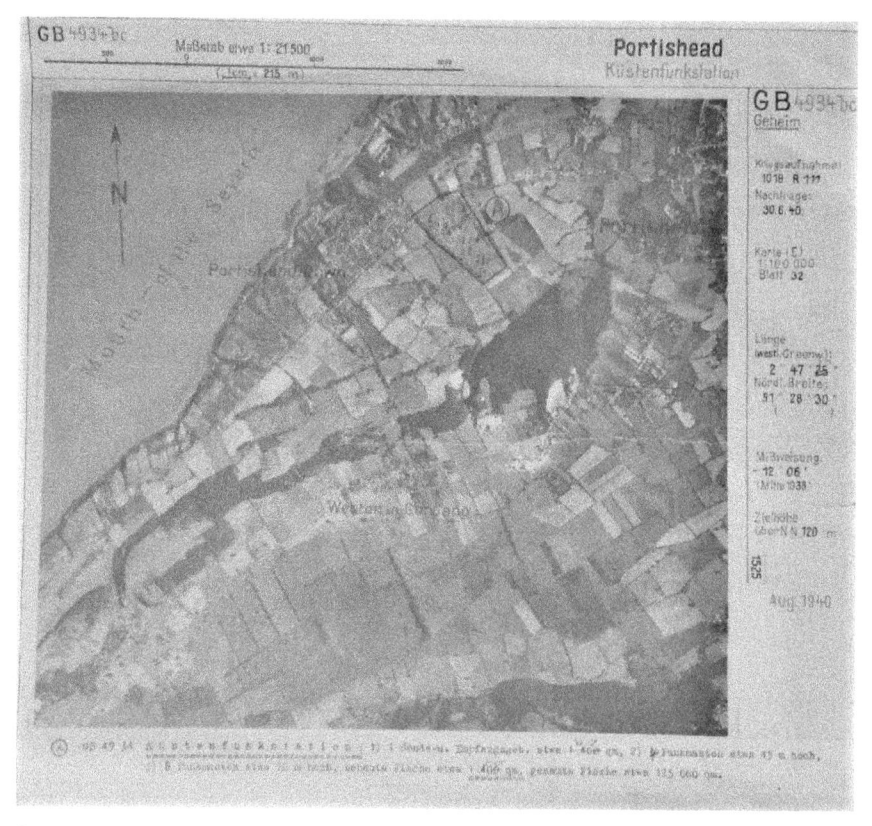

German reconnaissance photograph of the Portishead transmitter site from August 1940. The station is clearly identified.

In common with most buildings, blackout curtains and tape across the windows to reduce the impact of possible bomb blasts were installed, and remained in place until after the war.

In 1940, when anti-invasion measures were being taken, a network of emergency Inland Wireless Telegraphy stations was introduced all over the UK, which would provide an emergency inland wireless network for carrying urgent traffic in the event of the complete disruption of the inland telegraph and telephone service in any part of the country. These stations maintained a regular schedule with coast radio stations to ensure they were working satisfactorily, and were given radio call signs (similar to ships) to identify them, i.e.

EMERGENCY STATION	NORMAL COAST STATION	ALTERNATIVE COAST STATION
HEREFORD/GMBZ	BURNHAM/GRL	HUMBER/GKZ
LEEDS/GMDM	HUMBER/GKZ	CULLERCOATS/GCC
SOUTHAMPTON/GMDL	NITON/GNI	BURNHAM/GRL
INVERNESS/GMBV	WICK/GKR	HQ STATION/GMBB
BELFAST/GMBN	PORTPATRICK/GPK	SEAFORTH/GLV
HULL/GMDD	HUMBER/GKZ	CULLERCOATS/GCC
BRISTOL/GMFD	BURNHAM/GRL	NITON/GNI
CAMBRIDGE/GMBG	NORTH FORELAND/GNF	HUMBER/GKZ
ABERDEEN/GMCG	WICK/GKR	CULLERCOATS/GCC
BIRMINGHAM/GMCK	HUMBER/GKZ	BURNHAM/GRL
NOTTINGHAM/GMBD	HUMBER/GKZ	NORTH FORELAND/GNF
SHREWSBURY/GMBX	BURNHAM/GRL	SEAFORTH/GLV
CARLISLE/GMBQ	PORTPATRICK/GPK	CULLERCOATS/GCC
HARROGATE/GMBY	SEAFORTH/GLV	HUMBER/GKZ
EXETER/GMBR	BURNHAM/GRL	NITON/GNI
GLASGOW/GMCZ	PORTPATRICK/GPK	CULLERCOATS/GCC
DUNDEE/GMCM	WICK/GKR	CULLERCOATS/GCC
MANCHESTER/GMDG	SEAFORTH/GLV	HUMBER/GKZ
EDINBURGH/GMCV	WICK/GKR	CULLERCOATS/GCC
GLOUCESTER/GMBS	BURNHAM/GRL	NITON/GNI
READING/GMBM	BURNHAM/GRL	NORTH FORELAND/GNF
TUNBRIDGE WELLS/GMBP	NORTH FORELAND/GNF	HQ STATION/GMBB
CARDIFF/GMCL	BURNHAM/GRL	NITON/GNI
COLWYN BAY/GMBJ	SEAFORTH/GLV	PORTPATRICK/GPK

In addition, there were also smaller stations affiliated to the above, giving a total of 42 stations plus 11 mobile stations. These stations were manned by Post Office telegraphists, and located at secret locations in each of the above towns or cities. All stations had dedicated Medium Frequency and/or

High Frequency calling and working channels on which to contact their assigned Coast Radio stations. The scheme was periodically tested and remained in skeleton form during the rest of the war.

Recruitment of suitable staff for all stations became a problem, as a letter sent to all known officers, dated 31 March 1942 explains;

"On the 20th January 1941, a letter was addressed by the Postmaster-General to all suitable officers with experience of Morse working inviting them to offer their services for duty as Wireless Telegraphists for the duration of the war. The response was most satisfactory and large numbers of officers who then volunteered are now engaged at Wireless Stations throughout the country.

There is now, however, urgent need for additional volunteers to assists in the performance of important wireless communications at Post Office Wireless Stations in Scotland, Shropshire, and near London, and also at Post Office stations on the coast. All these stations are controlled, supervised and staffed entirely by Post Office personnel, and the work is connected with the safety of ships at sea, direction-finding, and the reception of wireless communications – work which forms a vital contribution to the war effort.

Ex-Morse Telegraphists who, after a refresher course, are able to send and receive up to 25 words per minute, but have not had experience of wireless working, would be given a suitable course of training in the duties they would be required to perform.

Those who have served in the Royal Navy, Army, Royal Air Force or Merchant Navy on wireless work and other officers (with or without teleprinter or typing qualifications) who have had some experience in the Morse reception of wireless messages would be particularly suitable.

It has been suggested that some officers who were unable to accept the Postmaster-General's invitation a year ago may welcome a further opportunity to volunteer and that, of others who accepted the former invitation, but later withdrew for domestic or other reasons, some may now be able to offer their services.

If you now wish to volunteer for this important work, perhaps you will kindly amend the enclosed option form and return it to me within seven days".

This obviously did not have the desired effect, so an internal Post Office memo, dated 20th May 1943, was sent to all Regional Directors regarding the recruitment of Temporary Wireless Telegraphists (female) which read;

"Applications are required from female Temporary Clerks grades II and III for temporary employment as Wireless Telegraphists at Post Office Wireless Stations in Scotland, Shropshire, and near London. Officers who are accepted for this duty will be regarded as on loan from their present Headquarters. They will be given training in Morse reception and typing in London prior to taking up duty at a Wireless Station. Previous experience in either of these subjects is not necessary. It is expected that employment on Wireless Telegraph duties will be available for at least 12 months. Officers who are registered under the National Service Acts and who have opted for 'The Forces' or 'Industry' may volunteer, and if accepted for the work the Ministry of Labour and National Service will arrange for their call up for the Forces or direction to Industry will be cancelled."

The weekly rates of pay for qualified staff were;

Age	Salary
At 18 years	43/-
At 19 years	47/-
At 20 years	51/-
At 21 years and over	55/-
Then by annual increments as follows;	
At 18 years	57/6
At 19 years	60/-
At 20 years	62/6
At 21 years and over	65/-
In addition to the foregoing rates, War Bonus is paid as follows;	
At 18 years	6/6
At 19 years	7/6
At 20 years	8/-
At 21 years and over	10/-

Around one thousand men and women were trained in Morse code and operating procedure as this type of work expanded, and sufficient ships' operators were examined to provide a continuous watch for ships at sea. In 1943, work had grown to such an extent that Naval operators from HMS Flowerdown were recruited to assist the civilian staff. In fact, many civilian operators were seconded to various Government services to man other radio stations and to train the many Radio Officers required for convoy work.

Life at Burnham and Portishead during the war seemed extremely busy, although security at both sites seemed quite relaxed, as Ernie Meaden recalls:

"Regarding our ground defences in WW2. The Army arrived almost immediately on the outbreak of war and remained throughout with 'lodgings' in one of our large Nissan Huts. Our staff of R/O's at the outbreak of the war was about 45 and we all had to join the Home Guard, doing relevant training now and then and taking a Sten gun with us on night duty. We saw very little of enemy activity at Burnham and Highbridge but there was one serious major incident when a bomb fell in Berrow Road and killed the wife of George Steer the station manager. The Navy arrived very quickly on the outbreak of war, 6 Petty Officer Telegraphists and a Lieutenant. Their numbers increased rapidly and they stayed for a very long time."

Don Mulholland also remembers:

"I know little about the Portishead site but imagine similar arrangements were taken as at the Burnham site. At Burnham there was a squad of infantry who performed sentry duty. There was little that could be done about such a large aerial site; it would have required a greater number of men to provide an adequate defence."

The Burnham Radio report of June 1942 advised that there was further activity in communication with aircraft due to increased night flying on the Atlantic route. 23 officers completed their training, some of whom had previously qualified for interception duties only. A Marconi CR100 receiver was tested and found to be very satisfactory. Delivery of 5 of these receivers followed.

However, a further report stated that 2 x CR100 receivers were installed on two points but in spite of the earlier favourable reports on a test model, it was the unanimous opinion of the manipulative staff (including Naval personnel) that the HRO receiver (in use at the time) was ahead of any other type.

Co-operation with Naval personnel continued in a satisfactory manner. Admiralty analysis of results obtained at Burnham revealed highly satisfactory performance and a much greater degree of efficiency of reception as compared with other Admiralty stations operating in the UK. The reasons ascribed to the report were:

(a) Specially selected staff

(b) Burnham's short-wave arrays

(c) General excellence of the receiving site

The sinking of the German battleship *'Scharnhorst'* was received (encrypted) at the radio station from HMS Duke of York and relayed to Admiralty as below:

Message transmitted 23.35 hours, 26 December 1943. Time of receipt at the Admiralty, via Burnham Radio Station was 02.42 hours, 27 December 1943. Text of message:

"My 262335 Part 1. After Force 1 had twice with their usual pugnacity driven and later shadowed SCHARNHORST, she was (finally?) brought to action and sunk at 1945A/ 26th December having been engaged by Force 1 and 2 and destroyers of convoy escort. The night action lasted 3.30 hours. At about 1830 it (seemed?) SCHARNHORST with her speed was escaping and in a very gallant action SAVAGE, SAUMAREZ, SCORPION and STORD N.S. succeeded in inflicting torpedo damage which enabled DUKE OF YORK to engage again. The main gun action between DUKE OF YORK (assisted?) by cruisers of Force 1 and (2?) and SCHARNHORST was prolonged, fought as it was in darkness and constantly increasing ranges. DUKE OF YORK (fired?) 45 rounds per gun. NORFOLK and SAUMAREZ slightly damaged. NORFOLK hit aft, casualties unknown (SAUMAREZ?) reduced to 8 knots, casualties 20. Some survivors (from?) SCHARNHORST (were?) picked up by our destroyers."

An interesting document sent from the Post Office Ship Inspection service to all UK Radio companies dated June 1941 gives full details of the Portishead Radio Short-Wave communication service in use at this time. The opening paragraph shows a little frustration at the inability of some vessels to maintain a regular contact;

"The wireless silence enforced on merchant ships since the opening of hostilities appears to have resulted, on occasion, in difficulty in establishing contact on short waves with Portishead Radio. The matter is viewed with some concern, and in order to improve the position, arrangements have been made whereby ships may, when permitted to do so by the Naval authorities, send practice messages to Portishead on short waves."

The scheduled hours of watchkeeping of Portishead Radio as published in official lists were modified as follows;

18 metres	Open for service from 0400 to 2000 GMT.
24 metres	Open continuously.
36 metres	Open continuously.
Ships should use the following frequencies for calling Portishead Radio:-	
18 metres	16560 kc/s (18.12m). GKS answers on 16845 kc/s (17.81m)
24 metres	12420 kc/s (24.15m). GKA answers on 12367 kc/s (24.26m)

| 36 metres | 8280 kc/s (36.23m). GKT answers on 8220 kc/s (36.50m) |

"Before leaving United Kingdom ports, ships equipped with short-wave apparatus should have their transmitters calibrated to each of the three calling frequencies (16560, 12420 and 8280 kc/s). Portishead Radio will assist, when possible, in this calibration. For this purpose, ships should call Portishead (GKT) on 8280 kc/s (36.23m) using the call sign GTST. Portishead will reply on 149 kc/s (2013m).

To enable ships when at sea to ascertain the most suitable frequency for calling, Portishead Radio emits the call signs GKS, GKA and GKT on the associated frequencies from 30 to 35 minutes past each hour. (The GKS call sign is emitted only during the watchkeeping period on this band - 0400 to 2000 GMT). The call sign which is best received will indicate the waveband on which communication could most probably most readily be established.

Ships transmitters should be adjusted hourly at -35 to the calling frequency in the waveband thus determined, and therefore be available for instant use in an emergency".

An updated notice was issued in November 1941;

"Arrangements have been modified and are now as follows;

Hours of watchkeeping;

18 Metres	Open for service from 0400 to 2000 GMT. Ships call, and are answered by GKS on 16845 kc/s (17.81m).
24 Metres	Open continuously. Ships call, and are answered, by GKQ on 12685 kc/s (23.65m).
36 Metres	Open continuously. Ships call, and are answered, by GKY on 8290 kc/s (36.19m).

The arrangements whereby ships were permitted under certain conditions to send messages for exercise have been cancelled.

Portishead Radio will assist as hitherto in the tuning of ships' transmitters to the new operating frequencies. Signalling incidental to tuning must be kept to a minimum, and permission of the Naval authorities obtained beforehand".

It would appear from other associated documents that the war was nothing more than a minor irritation for the Ship Inspection service; expense claims, stationery orders and travel documentation continued as if nothing had happened.

All British Coast stations were issued with official documentation with regard to wartime processes, involving such items as intercepting traffic from foreign coast stations, the set-up of reserve and emergency stations, air raid precautions, and various security-related procedures. Many of these were obviously of a common-sense nature, but other instructions, especially to do with the interception of traffic, were somewhat involved. Basically, these instructions gave details of the type of traffic which would be of interest to the UK authorities, and the type of codes being used. For example, "Traffic consisting of German nouns and proper names grouped together in an apparently meaningless fashion is urgently required".

The 'Passau Interception Service' also required details of coded messages transmitted by the German authorities on various medium and short-wave frequencies, although the need for these ceased on 17th May 1945.

As many 'new' wireless telegraphists were being employed at Post Office coast radio stations, a guidance booklet was issued to each person giving details of communication procedures, log-keeping and the various 'Q' and 'Z' codes commonly in use, together with a basic guide to radio propagation.

Signalling both to and from ships during the wartime period was kept to a minimum. To ensure this, a new method of message handling was needed and subsequently adopted; this method became the fore-runner of the Commonwealth 'Area scheme'. By necessity, ships' call signs had to be changed to secret 'wartime' call signs, which helped to protect the identity of each vessel. A Radio Officer of the time, Robert Veitch, notes;

"During the war, I had on one occasion to report the shelling of the motor tanker *'Empire Gold'* by a surfaced submarine. The vessel was on passage to Galveston Texas and was approaching the Florida Straits. Neither Bermuda Radio nor Trinidad Radio repeated the distress message. In the circumstances I obtained permission from our captain to communicate the details to Portishead using a secret war call sign hoping of course, that our position would not be disclosed. Fortunately, Portishead answered first call and took the details, subsequently broadcast by Bermuda Radio. The *'Empire Gold'* arrived in Galveston a few days ahead of us albeit with a large shell hole in the hull".

As mentioned above, it was necessary to ensure that a message for a ship would be received on the first transmission by the coast station, and that a ship requiring to send a message should be able to establish contact, ideally with a single call. Ships would only make contact in case of emergency, but it was imperative that they were answered immediately. For this reason, it was decided to keep radio watch for ships on common frequencies at a number of Commonwealth stations; each station was equipped to reply on the same frequencies. By this means, a ship heard calling any one of these stations on any one of these frequencies would be answered, if not immediately by the station being called, by any one of the other stations in the network. Messages for ships were sent to the nearest Commonwealth station and broadcast at fixed scheduled times.

Radio Officer Nigel Williams, at that time serving as 3rd R/O on the *'Highland Prince'* recalls the period well;

"During the war, all ships listened to the BAMS (British and Allied Merchant Shipping) broadcasts of the traffic lists and traffic every four hours for the area they were in. The whole world was divided into BAMS areas e.g. BAMS1, BAMS2 etc., and local coast stations transmitted the traffic for their area. The broadcasts began with a list of the secret call signs of the ships for which they had traffic, and then the messages themselves were broadcast. Each message was repeated four hours later at the next BAMS broadcast. I seem to remember that Portishead Radio also transmitted the BAMS traffic, as did Rugby Radio (GBR) on 16 kHz.

When in convoy, only certain ships would be asked to keep continuous listening watch on 500 kHz (international distress and calling frequency). No doubt the Senior Officer Escort (usually a frigate or a sloop) would do the same, and would in an emergency us 2182 kHz AM to talk to the other escort ships, usually corvettes, although normal signalling was by lamp or flag. This means that in convoy, there was nothing for the R/Os to do apart from reading the BAMS traffic lists, so we would spend our watches on the bridge, and became proficient signallers with the Aldis lamp and with signal flags. Wireless silence was observed at all times.

When sailing as an independent ship (i.e. not in convoy), continuous listening watch would be maintained on 500 kHz, and wireless silence would only be broken in an emergency or if attacked. The following three distress signals were used if attacked;

- SSSS sent three times, followed by the ship's wartime call sign and position, if attacked by a submarine.
- AAAA (as above) if attacked by aircraft.
- RRRR (as above) if attacked by surface raider.

These signals would be sent on 500 kHz. I am glad to say I never had to send any of the above distress signals, but I did break wireless silence once in about 1944 when I was chief R/O on a collier in the Indian Ocean and our bunkers went on fire. The old man (captain) gave me the message to code and recode and, following the standard procedure, I called GZZ (Whitehall) on HF and passed the message to the first HF coast station to reply, which I think was VIS (Sydney). All these stations were linked by a world-wide teleprinter network, and it was said that any signal would be in Whitehall within ten minutes. I then waited for our wartime call sign to appear in the

BAMS broadcast, copied the message to us, decoded it, and gave it to the old man".

Radio Officer John Edwards also recalls wartime radio communications with Portishead;

"During my wartime experience, there were two anchor points back in the U.K., wherever in the world one happened to be - Rugby Radio and Portishead Radio. The former was listened to every day for the Greenwich time signal, essential for navigation purposes, and the latter for totally 'blind' traffic lists consisting of individual ships' call signs and collective 'GBMS' (Great Britain Merchant Ships), soon changed to 'BAMS' (see above). There were also a few collective operating company messages, but never was a 'live' ship's Morse key touched to QSL (acknowledge receipt) - if you missed your message, bad luck - but keep quiet, whoever you were! Apart from the normal published lists, each ship also had a 'secret' call sign, which one gathers from definitive post-war narratives, was about as secret as the colour of the funnel!"

Another Radio Officer, Phil Mitchell, recollects;

"I was at sea with Marconi from 1937-1945, and do remember at on the few ships equipped with HF Radio, GKT (as it was then universally known) had a note that was instantly recognisable because of its characteristic note; it sounded more like a spark transmitter than some of its competitors on the continent. At the outbreak of war, I believe Burnham took over the keying of the VLF Rugby transmitter GBR, which together with its counterpart in the Falkland Islands provided world-wide broadcast coverage for shipping until the Area scheme got under way. Ships were required to keep radio silence, and to be dependent on QSA1 signals (weak signal strength) from GBR at 0000 and 1200 GMT in the middle of the Pacific and a tropical electric storm was quite a responsibility".

Radio Officer Bill Hall served on board HMML 349 Eastern Mediterranean during 1942 and 1943. He recalls;

"Here comes my first association with Portishead Radio and that band of efficient and dedicated operators. A transmission from my little low powered transmitter, with its purloined HF attachment from the coast of the North African desert, never failed to be received by Portishead Radio. There was however one slight problem and that was the performance of the receiver on Portishead's transmitting frequency. The receiver was capable of receiving Alexandria Radio on their high power LF broadcast transmitter, and on instructing Portishead, the reply would be broadcast from Alexandria Radio. It never failed, and on one occasion I was pleased to receive the following message;

HMML FROM C IN C MEDITERRANEAN

ML349 IS TO BE CONGRATULATED ON HIS EFFICIENT WIRELESS COMMUNICATIONS.

I did not get any uplift in my pay but it was nice to receive the signal. The real congratulations should have of course gone to the operators at Portishead Radio."

On moving to a destroyer in 1944 in the Mediterranean, Portishead was still my mainstay of communication and I could not praise them enough".

The area system proved to be so successful that its use was continued after the cessation of hostilities, and many of the Naval staff employed at Portishead were retained to work alongside their civilian colleagues. Their presence continued at Portishead until 1971, when the last Naval Radio Officer left the service.

Blackout curtains still in place at a landline position just after the war.

Recently-unearthed annual reports on the Wireless Telegraphy section of the Post Office give an excellent breakdown on the amount of traffic being handled not only by the smaller MF coast stations, but also at Portishead Radio.

For example, between September 1943 and September 1944, 727 distresses were handled by the section, with more aircraft distresses being dealt with. Broadcasts to ships made on behalf of the Admiralty amounted to about 152,000 code words, making a total for the war period of well over

4,000,000. Traffic intercepted for the use of Naval intelligence amounted to over 2,500,000 words, bringing the total for the five years to nearly thirteen million.

Civil aircraft traffic handled at Burnham remained constant at about 50,000 words per month. This traffic was mainly with aircraft flying to and from Lisbon, Lagos, and Bathurst. Broadcasts to Lisbon for the information of the British Air Attaché were also made, and interceptions from Lisbon for the information of the Air Ministry. Traffic too was exchanged with air-ground stations at Alexandria, Asmara, Bathurst, Lagos, and Tripoli.

Records remain of all distress cases handled by all UK Coast Stations (including Burnham/Portishead), and from the figures, it is clear how vital the stations were during the war years in ensuring safe passages of both merchant and Naval vessels.

QUARTER	SHIP DISTRESS CASES	AIRCRAFT DISTRESS CASES	TOTAL
SEP-NOV 1939	217	22	239
DEC 1939-FEB 1940	283	25	308
MAR-MAY 1940	153	49	202
JUN-AUG 1940	236	52	288
SEP-NOV 1940	247	27	274
DEC 1940-FEB 1941	334	23	357
MAR-MAY 1941	269	23	292
JUN-AUG 1941	138	54	192
SEP-NOV 1941	108	27	135
DEC 1941-FEB 1942	236	22	258
MAR-MAY 1942	191	68	259
JUN-AUG 1942	89	64	153

SEP-NOV 1942	199	33	232
DEC 1942-FEB 1943	199	64	263
MAR-MAY 1943	163	84	247
JUN-AUG 1943	54	69	123
SEP-NOV 1943	85	98	183
DEC 1943-FEB 1944	114	86	200
MAR-MAY 1944	77	79	156
JUN-AUG 1944	94	94	188
5-YEAR TOTAL	**3486**	**1063**	**4549**

The maritime distress cases include attacks by submarines, aircraft and surface raiders, and also cover distresses not caused by enemy action. However, it is clear that with over 4,500 incidents handled, the operators at the Coast Stations were not kept idle.

Upon the resumption of commercial working after the war, the tariff rates were increased to 1/- per word (full rate), 6d per word (reduced rate) and 4d per word (SLT - minimum 6/8d for twenty words). A charge of 6d per word was fixed for the broadcast of weather forecasts and other navigational warnings.

Less well documented is the role which the station had in contacting allied 'special' radio stations in Europe. It is common knowledge that the station helped in the reception of messages from allied agents in Europe and Special Operations Executive (SOE) stations, but details are hard to trace. However, correspondence received from Svein Korsvold in Norway sheds light on one particular episode involving his father and one Thomas William Murray:

"He was born in Preston on April 10th, 1889 and was supposed to have been a wireless instructor at the Curragh camp at a very young age. He joined the Marconi Company in 1911 according to the Marconi archive. He worked for them until 1920, and again from 1947 to his death in 1949. He was 1st telegraphist during the Great War on many different ships, but often White Star Line. He served on the *'Empress of Britain'* during her days as a troopship for the Dardanelles campaign (1915).

After moving to the US with his family he started working for the Federal Government. Eventually at least 5 years undercover in the mafia and was a key witness in the trial against mafia boss Johnny Torrio in April 1939. Soon after he joined the Norwegian ship *'Songa'* as Radio Officer.

On January 22nd, 1940 the *'Songa'* was torpedoed outside Ireland. The crew survived, and he eventually ended up in Bergen, Norway together with the rest. He checked in to the Rosenkranz hotel on April 6th.

On April 9th the Germans invaded. My father and some of his friends started a resistance group, and Thomas William Murray joined them and helped to get in contact with the UK. They had radio amateurs among them, had no luck until Murray took over. He got through immediately. They used the call sign LMA, and sent to GKD/GKC on 24m according to a report from one of the radio amateurs working with Murray. This made them the first station to make contact with the UK from occupied Norway.

He escaped to the UK on the Fishing Vessel *'Sjøglimt'*, arriving at Lerwick on June 2nd, together with (amongst others) Commander Villiers from the British convoy office in Bergen. In 1941 he was headhunted by the MI6, and stayed there for the rest of the war".

In the middle of August 1941, the Naval monitoring station at Portishead began to pick up weak, intermittent signals from an unidentified wireless transmitter which purported to be working inside Yugoslavia. The contact was carefully nursed, and by the end of the month it had been established beyond doubt that the station was being operated by units of the Royal Yugoslav Army which were still offering vigorous and well organised military resistance to the enemy in the hills of western Serbia. The headquarters of these forces were on Ravna Gora, some twenty miles due north of Čačak, and they were under the command of a Colonel Draža Mihailović, the senior officer present. The call sign 'Bullseye' was used for this station.

The same operators maintained their schedule so that any other different operators at the other end would be instantly recognised by their different methods of hand keying.

There was a story too that the 'X' service (GCHQ) could have been based at the Highbridge site. Some of the Portishead R/Os transferred temporarily to the nearby 'X' service station at Alstone (Huntspill), where an associated one-man operated Direction Finding (D.F.) hut was located in a field towards the sea wall.

On the cessation of hostilities in 1945, the station geared itself up for some dramatic changes.

CHAPTER THREE

POST-WAR AND THE 1950S

In 1945, it was reported that the partial resumption of commercial working, consequent on the end of the European war, had little effect on the station. The removal of restrictions from private radiotelegrams was, however, keenly anticipated in order that the stagnant state of the HF Ship-to-Shore service would come to an end. Activity in the Civil Aviation section remained brisk with average traffic figures of 25,000 words weekly.

In fact, it was found that traffic totals in the first six months of 1947 showed an increase of approximately 80% above the totals for the corresponding period in 1946.

An address given to the Radiocommunication Convention by Colonel Sir A. Stanley Angwin on 25th March 1947 gives a concise overview of the role of the long-distance ship-to-shore service during the conflict;

"The changes brought about by the war in the long-distance telegraph service to ships have been mainly in methods of operation, and changes in equipment have been largely incidental to those operating changes. Before the war, traffic to ships originating in this country, or from ships for delivery in this country, was handled by the twin radio stations at Burnham and Portishead in Somerset; Burnham is the receiving station, the transmitters being situated at Portishead and operated by remote control from Burnham. Messages for ships handed in at Post Offices were sent to Burnham and at scheduled times a 'traffic list' was transmitted. This traffic list consisted of the call signs of all ships for which messages were on hand. A ship's operator, hearing his call sign in the traffic list, would call Burnham, and, contact having been established, the message would be passed on to the ship. Messages from the ship to Burnham were dealt with similarly.

With the outbreak of war, Burnham and Portishead, as well as the coast stations, were taken over by the Admiralty. The Post Office continued to staff and manage the stations, but operationally they became Admiralty stations. A small number of Naval telegraphists were stationed at Burnham to augment the Post Office staff. It was important that ships at sea should not disclose their positions to the enemy, so that all radio signalling had to be reduced to a minimum. The world was divided into areas, in each of which was situated an area transmitting and receiving station (in some cases the receiving stations were duplicated). Messages for ships were broadcast from the appropriate area stations at scheduled times, but no acknowledgement of receipt was given by the ships. A network of point-to-point communication channels was set up between the area stations, so that messages originating anywhere in the world could be routed to the appropriate area stations.

In some cases, it was essential for ships to send messages to shore, and in order to avoid repetitions by the ship (increasing the risk of betraying the ship's position to the enemy) duplicate receiving stations were set up in some areas. In this country, such a station was established in the West Midlands. This was equipped with a set of directional aerials for receiving from all quarters of the globe. The operators were in touch, by means of loud-speaking telephones, with their colleagues in Burnham. It was often found that in conditions of fading or interference parts of a message missed at one station would be received clearly at the other, and the ship was not required to repeat the message.

The number of Radio Officers carried in merchant ships was nearly trebled to ensure a continuous operator watch at sea. The expansion is shown by the fact that, whereas in 1938 the numbers of certificates of proficiency in radio operating issued by the Post Office totalled 531, during the war an average of 2,500 per year were issued".

At the end of the war, it was decided to retain the broadcast method of passing traffic to British ships and to ships of such foreign countries as wished to take part in the scheme. Traffic for other ships was handled by pre-war methods. A message for a British ship, handed in at a Post Office in this country, was sent by teleprinter to Burnham, where a Ships Bureau was maintained. Records of the position and course of all ships were kept at the Bureau. The message was sent on over point-to-point channels to the appropriate area station. Ships notified Burnham Radio when they entered or left port and when they went from one area to the next, so that the records in the Ships Bureau would be kept up to date. This long-distance ship service, evolved as a war requirement and now expanded, proved most efficient for normal ship-shore communication.

With the coming of peace and the resumption of normal commercial working, it became necessary to re-organise the services provided by Portishead Radio. Despite the restrictions imposed by war work, plans were being made as early as 1944 to improve the service, which included the expanding of Portishead Radio with new purpose-built extensions, the installation of extra consoles to handle the anticipated increase in traffic, and new traffic handling procedures.

A description of the work at the UK coast radio stations, published in 1945, today makes interesting reading;

The Area Broadcast room in 1946. Area 1A broadcast position on the left with a Naval telegraphist on the right sending received Naval signals on the direct line to the Admiralty.

"Wireless Telegraphists are normally employed at nine Coast stations, one special service station (St. Albans), and on two cable ships. They are liable for duty at any station.

Service aboard the cable ships is compulsory on staff entering the service in 1929 and later. The duties aboard the cable ships are rotated among the whole of the wireless staff and have been filled by applicants volunteering for two-year spells. The operator's work aboard these ships is similar in character to that of the normal duties of the seagoing wireless operator.

The following is a list of general duties performed by Wireless Telegraphists:

(a) Short-wave watch keeping and traffic working.

(b) Medium-wave watch keeping and traffic working.

(c) Distress watch keeping and traffic working

(d) Direction-finding

(e) Radio Telephony and small craft distress and traffic working

(f) Creed perforating

(g) Slip writing and typing

(h) High-speed recording

(i) Landline telegraphy

(j) Maintenance Duties

(k) Ships Bureau

(l) Accounting

(m) Writing Duty

(n) General"

The article goes on to describe the operation of a Portishead W/T Search Point;

"Searching Points are staffed by operators on each wave advertised as open for communication. The search operator, wearing headphones continually, varies a tuning condenser up and down the limits of the wave band allotted. In this wave band, ships of all nationalities may call and work to each other, or call any land station. Dozens of ships are audible. It is the duty of the Search Operator to search and 'pick up' a station calling him. Each station has call letters allotted. Having identified the ship requiring to communicate, the Search Operator logs the signal and informs a Traffic Operator on an adjacent set as to the identity and approximate position on the wave band. The Traffic Operator then calls and picks up the ship on his own receiver and transfers to a 'working' wave.

This operation calls for skill and speed. Having once established contact, the Traffic Operator reads Morse signals of varying speeds and strengths through all types of interference caused by other ships, induction, and atmospherics. A click of an atmospheric can easily change the character of a letter in a work, and it can be readily understood what this may mean in a code message. As an example, the letter O (- - -) may be split up by an atmospheric into the letter X (-. . -). These pitfalls are only avoided by skill and intense concentration, with a genuine interest in the work.

Short-wave signals require very 'fine' tuning. Short waves are used both for commercial traffic working and for distress signals"

In addition, the Portishead Radio Officer had to use his skills for Landline work;

"All stations are connected by landline to the nearest town. Burnham is connected to the Central Telegraph Office (CTO) London, and the Admiralty, London. Both Teleprinter and Telex are used and normal landline practice carried out. All Wireless Telegraphists are fully proficient in touch-typing but are able to revert to Morse Sounder working in the event of a landline breakdown".

Use was also being made of the German Coast Radio stations (especially those located at Kiel and Wilhelmshaven), to provide MF services to ships under Allied control, although still manned by German operators. The Norddeich Radio HF was not allowed to provide HF services to ships but its transmitters were being used for other Naval purposes.

After the war, it was apparently traditional for a member of the radio station staff (although on one occasion an R/O at Land's End/GLD) to sail on the maiden transatlantic voyage of the Cunard liners. However, it is not known when this most welcome invitation was rescinded. From January 1st, 1946, British and Commonwealth ships used an enhanced version of the 'Area Scheme' developed during the war. In this new scheme, now known as the 'British Commonwealth Area Scheme', the world was divided into a number of areas, each having its area transmitting and receiving station. Traffic was sent along Admiralty routes to the station in whose area the ship was sailing, and then broadcast at four-hourly intervals; 0000, 0400, 0800, etc. Acknowledgement by the ship was given as soon as possible after receipt. Ships would notify changes of area and position to any station in the scheme and these details passed to Portishead Radio, where a 'Ships Bureau' maintained an up to date card index of all ships' positions.

Consideration was also given to allowing USA-flag vessels to use the scheme, but this proposal was shelved when the American authorities would not be prepared to make a 'substantial' contribution to the scheme by providing 2 or 3 new Area Stations. It was also felt that the much-increased workload which would result from these new stations would cause problems at Burnham. The whole scheme was outlined in a 6-page detailed proposal dated 5th July 1945, and relevant extracts are quoted below;

View of the upstairs receiving room at Burnham c. 1946. A Naval R/O clearly visible.

"The stations concerned overseas will include (subject to the degree of participation being agreed in certain cases) Simonstown, Falklands, Colombo and certain stations in the Pacific area together with Admiralty stations at Gibraltar, Malta, Alexandria, Bermuda, etc., which may be keeping watch on ships frequencies for the reception of TRs, acknowledgements, and traffic.

Traffic for Merchant Ships is dealt with as follows;

a) Messages which can be cleared by short-range coast stations are routed by Burnham and forwarded to the appropriate station.

b) Messages for long distance are subdivided as follows:-

1. For British ships, and foreign ships participating in the scheme

2. For Foreign ships not participating.

Messages in (1) above are again subdivided and routed for Area I, Area II, Area III, Area IV and Pacific Area.

Messages originating overseas and received over point-to-point links via Admiralty for Area I will be included with Area I traffic; those received from overseas for routeing will be routed to Burnham to the appropriate area and returned to Admiralty.

Operating area c.1946 with Radio Officers Pharo, Miller, Wynne and Waugh in action.

Area I traffic will be prepared for broadcast and transmitted by Burnham at the appropriate times and on suitable frequencies. Area II, III and IV traffic will be routed and sent to Admiralty for passing over to the appropriate Area station for broadcast. Pacific traffic (until other arrangements are made) and traffic for non-participating foreign ships will be passed to the 'traffic point' and held for clearance under pre-war (direct working) arrangements as will traffic for any ships for which special North Atlantic arrangements have been made.

TRs and ship's positions received either direct or via short-range coast stations or via Admiralty will be passed to TR point for recording (and plotting if incorporated). Navigation warnings and meteorological messages will

Transmitter requirements are as follows:-

1) For Broadcast.

Burnham will require control of a long wave transmitter together with two short-wave transmitters every four hours for a two hour period, covering Area I and Area IV (probably). Appropriate arrays will require to be associated with short-wave transmitters. This arrangement covers British and foreign participating ships in Areas I and IV. Timings of schedules will need to be co-ordinated with those required by Admiralty for Admiralty traffic to H.M. Ships.

2) For Answering.

For answering ships calling Portishead Radio with acknowledgements, TRs and messages, four short-wave transmitters will be needed;

One transmitter in 4 Mc/s band – 2200 to 1000 hrs GMT

One transmitter in 6 Mc/s band – continuous.

One transmitter in 8 Mc/s band – continuous.

One transmitter in 12 Mc/s band – 0800 to 2400 hrs GMT

One transmitter in 16 Mc/s band – 1000 to 2200 hrs GMT

These transmitters cover British, foreign and H.M. ships throughout the world (long-distance) as well as any North Atlantic ships for which special arrangements have been made.

3) For traffic by direct working.

For handling traffic from Burnham by direct working to British, foreign and H.M. ships in Pacific area (if other arrangements have not been made) and to North Atlantic ships for which special arrangements have been made, four short-wave transmitters corresponding in frequencies and hours of working to the four answering transmitters will be needed. These transmitters cover also foreign non-participating ships in Areas I, II, III and IV.

In addition, two medium-wave transmitters – answering and traffic – on 143 and 149 kc/s are required for foreign and North Atlantic (special arrangement) British merchant ships. They will need to be continuously available.

Transmitters referred to above should as far as possible be Portishead transmitters to avoid long control lines.

Receiving station arrangements require to cover the reception of TRs, acknowledgements and traffic (as applicable) from H.M. Ships, British and foreign merchant ships participating in the scheme and foreign ships not participating. They need to envisage reception on high frequencies and, in the case of Burnham, medium frequencies in the 110-160 kc/s band as well as high frequencies. Arrangements for communications in both cases are similar to the extent that calls from ships are made on calling frequencies. TRs and acknowledgements may be passed on calling frequencies but traffic must only be passed on traffic frequencies to which ships (and shore stations where this applies) must change if initial contact has been established on calling frequencies.

Arrangements for foreign ships not participating should not cause any great difficulty. Calls and from-ship traffic can be accepted at Burnham under the scheme and to-ship traffic can be passed from Portishead by traffic points on traffic frequencies. The only variations necessary seen to be those requiring traffic lists to be notified and made especially for these ships, and the listing at receiver points of nationalities not participating".

The document goes on to discuss receiving procedure (in great depth), receiving station operating layout, details of the traffic bureau and circulation centre, reception equipment, and transmitter control equipment.

The Portishead transmitter site used Marconi SWB11 transmitters for the Area 1 section, with traffic lists being sent every 4 hours. The first commercial voyage of the liner *'Queen Elizabeth'* during the year proved exceptionally busy for the station, with over 66,000 words of traffic sent to the vessel.

The resumption in service to merchant vessels was extensively advertised, with newspaper reports happy to publicise the new procedures:

"The radio telegram service to merchant ships in all parts of the world will be resumed on January 1, 1946, and the war experience will be applied to peacetime long-distance communication. Capetown has been selected as the transmitting and receiving station for Area 2, defined as that part of the Atlantic east of 30 degrees west and south of the equator, and that part of the Indian Ocean west of 45 degrees east and south of 18 degrees. The powerful apparatus available for Capetown Radio will enable communication with vessels in any part of the world. Radio telegrams will cost 1s a word plain, and 7d code."

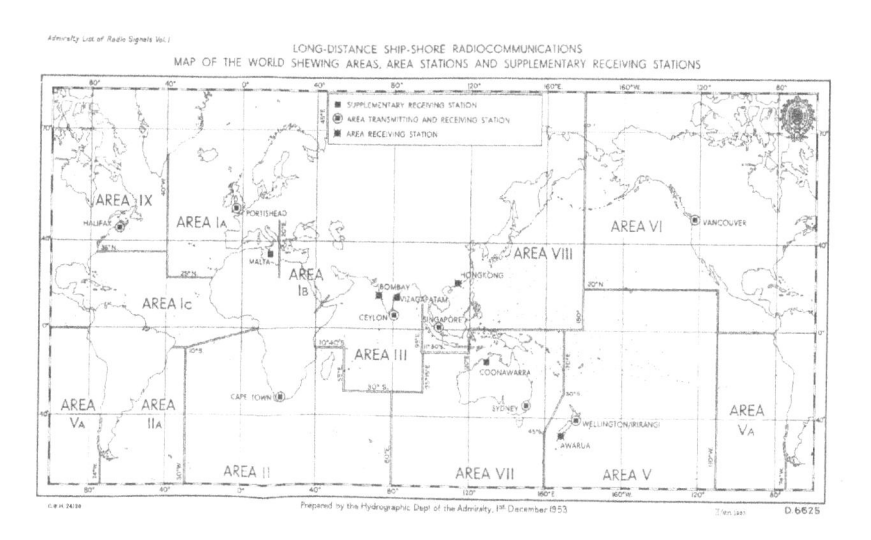

The 'Area Scheme'. Although this map was produced in 1953, the areas remained valid for the duration of the scheme.

This 'Ship's Bureau' position was at the time an extremely important aspect of Portishead's work. However, the actual task of locating the Bureau at Portishead was not without its problems; there was pressure on the Post Office to maintain the work at the London Central Telegraph Office (CTO), and only a stern inter-departmental memo dated January 1946 made sure the transfer took place. Extracts from this memo read;

"It is realised that the CTO is better placed geographically to deal with London enquiries; but Burnham is not less well placed as regards the country as a whole. All the relevant features of the new scheme (area scheme) have been carefully weighed and it is considered that, on balance, centralisation at Burnham is the most advantageous course. The arrangements for the siting of the Ships Bureau at Burnham have been reviewed and it has been decided that they must go ahead"

The map of the world is divided into nine areas. Ships communicate their position every two days. From this information, or information obtained from various shipping lists, plus estimated speed and course the Bureau operator indicates the area station to which the message is sent. A good geographical sense, knowledge of ports and shipping routes is essential for performing this type of work.

Foreign ships do not come under the area routeing scheme; they are worked on a direct contact system and routeing takes a different form.

The 'Ships Bureau' at the old upstairs station at Burnham in 1946. TRs were filed in the cabinet on the left. The GRL console was located through the door at the top of the stairs, looking outwards to the front gate.

If the ship is fitted with short-wave and estimated to be in 'skip' distance, that is, too close in to communicate on certain wavelengths, the message is forwarded to a coast station under a mutual clearance advice service.

The Bureau operator's knowledge of shipping routes, estimation of speed and course plays an important part in determining the redirection of messages to the flank coast stations.

Service messages relating to non-delivery, interim advices of transmission and the usual official enquiry messages are dealt with by the Bureau operator.

'TRs' (messages received from ships indicating their position, next port of call, etc.) are entered and filed at the Bureau position".

In the ship-to-shore direction, traffic was normally sent to the area station nearest to the country of destination of the message, but in cases of communication difficulty it could be sent (without extra charge) to any other area station. Communication between area stations was carried out over the Admiralty point-to-point radio network.

The rate being charged for such messages were 1/- per word for radiotelegrams and 6/8d for Ship Letter Telegrams (SLTs) for 20 words, regardless of routeing. Charges accrued to the station to which the radiotelegram was addressed, and redirection over the Admiralty Network was free of charge.

Foreign-flag vessels (i.e. those vessels registered in non-British or Commonwealth countries) continued to use the 'direct working' system introduced prior to the war in both to-ship and from-ship directions. However, those Radio Officers serving on British-flag vessels enjoyed the facility of contacting their local coast station rather than seek a propagation 'window' to contact Portishead directly. Radio Officer Nigel Williams was serving on the *'Queen Elizabeth'* at the end of the war, and remembers setting up a radiotelephone link through the Baldock Radio radiotelephone installation;

"On Christmas Day 1945 the BBC decided they would like to link up with the *'Queen Elizabeth'* in mid-Atlantic, prior to the Queen's speech at 3.00 p.m. The idea was to exchange greetings with the captain and others on board, and to finish off with 'the ship saying farewell' by sounding her siren. Just in case contact could not be made on the day, a party of BBC engineers came aboard at Southampton before we sailed, complete with recording equipment, and I suppose must have recorded the whole proposed programme; the only part that I assisted with was the recording in the wheelhouse of the sound of the siren, and I well remember the BBC man calling for a towel to wrap around his microphone once he had been treated to the sound!

On the day everything went according to plan and the recording was not used, although at the time, if I remember correctly, we were in the teeth of an Atlantic gale and were running at reduced speed".

He continues;

"After the *'Queen Elizabeth'* and *'Queen Mary'* were refitted for peacetime use, I spent a short time with them. The telegraph traffic that we exchanged with Portishead (mostly GKU on the long wave) reached a possible all-time peak at that time. Many a time it was a question of 'QTC 10 AS MO' (I have 10 telegrams please wait a moment) as one operator slid out of the chair and another slid in!"

It is interesting to observe that traffic levels both to and from the vessel were extremely high - 49,078 words from-ship, and 17,138 words to-ship. This was indeed well in excess of the load handled during the maiden voyage of the *'Queen Mary'*. The automatic transmission in the above telegram refers to the equipment at use at Portishead which could send telegraphy messages at high speed and decoded with equipment on board the vessel. Prior to the transatlantic crossing, a trial trip was made from Southampton to Greenock on 7th October, to test this equipment. A Post Office operator from Portishead was on board for this trial, and reported that the tests had resulted in speeds of 100 words per minute on the long wave circuit between the ship and Burnham."

Phil Mitchell remembers this time well, and casts an operator's view on the high-speed system mentioned above;

"When *'QE'* (Queen Elizabeth) and *'QM'* (Queen Mary) returned to full service after the war, Burnham moved down from the original operating room (upstairs) into the then-new building. In what was seen (then) as a highly technical development, four operating positions in 'C' wing had been equipped with a facility to take high-speed Morse (up to around 120 wpm (I guess). It was expected that ships would fit compatible facilities to allow them to clear traffic (and press in the case of the *'QE'* in less air-time). On *'QE'*s first post-war trip we copied several yards of newsworthy tape before it was realised that we only had a few operators able to read from the tape. We also found that it took as long (if not longer) to translate the tape before it could be sent down the line, so we promptly reverted to normal operating procedures. And so far as I can recollect the special facilities were never used in anger again. It took a long time to persuade our engineering friends that our main business concern was how to take a little traffic from a lot of ships, not vice-versa, and that the S.S. *'Cesspool'* was most unlikely to need press facilities or for that matter to become a telephone box".

Even though high traffic vessels like the *'Queen Elizabeth'* were (supposedly) making use of the automatic transmission/reception facility, the vast majority of ships still needed to make use of the manual method of telegraphy communication.

As more ships became fitted with high-frequency radio equipment, some 'fine-tuning' of operational procedures was needed, an example of which was noted in a draft progress report in the summer of 1947;

"The service continues to operate satisfactorily but some difficulty is being experienced between the hours of noon and midnight is disposing of the heavy 'To-Ship' traffic during the 1-1½ periods allocated for the transmission of radiotelegrams to ships.

The question was discussed at two ship-to-shore working party meetings particularly with regard to handling the heavy traffic to be expected at Christmas. It was finally agreed, as an experimental measure, to divide the large area covered by Portishead into three sub-areas, thus enabling separate (but simultaneous) transmissions of traffic to take place on different frequencies for each sub-area. During periods of light traffic however, the three sub-areas would be combined and traffic transmitted as present in one transmission.

The scheme will be brought into operation on the 1st November 1947, so that experience may be gained of its operation before the Christmas traffic starts to build up".

As with most 'temporary' ideas within the service, this was to become a permanent arrangement – areas 1A, 1B and 1C were brought into full usage during 1948. Staff problems were also noted;

"Shortage of staff at Burnham is causing some anxiety. Recruitment has been very much below expectations and efforts are now being made to obtain acting radio operators from the inland service".

An article written by Portishead Radio's officer-in-charge Mr. W. Swanson in the November/December 1948 issue of the 'Marconi Mariner' confirms the procedures used for Wireless Telegraphy operation at that time;

"Two methods of dealing with calls from ships on high frequency are in operation at Portishead Radio. The first method is used on all frequencies during periods when the number of ships to be worked in any one frequency band does not require more than two operators. Each operator searches over and beyond the ships' calling band; he answers ships as they are heard, and moves them outside the calling band as necessary for reception of traffic. In general this method is used in the 6 M/cs band throughout the 24 hours, and in the 8 M/cs band from 0000 to 1600 GMT.

The second method requires the employment in each frequency band of an operator who continuously searches for ships' calls; it is in operation as follows:

16 M/cs.	0800 - 1800 GMT.
12 M/cs.	0800 - 2000 GMT.
8 M/cs.	1600 - 2400 GMT.

Each search point operator may, according to traffic conditions, have from two to five 'workers' associated with him.

Every receiver is calibrated at the centre of each of the ships' calling bands, and receiver dial readings are exhibited at all operating points. Each operating point has facilities to control any of the HF transmitters; but each transmitter is shared by all operators allocated to a particular frequency band. When, for example, an operator is using GKS the 'send-receive' switch on his Morse key operates a red warning light at the other 16 M/cs receiving points.

Search point operators are changed every two hours. They keep a special log; in it they record the time of receipt of calls, ships' call-signs, receiver dial readings in degrees above or below the centre of the band, the best aerial for reception from a particular ship, and the time of transferring a call to a working point operator. If all working operators are engaged, they give ships a turn (or QRY); these ships are then passed out in strict order as working points become disengaged.

A working operator, on being 'handed out' a ship, makes a rapid swing above and below the centre of the band to determine the channel most free from interference. He then moves the ship up or down so many k/cs for the reception of traffic.

Experience has shown that the search point method is the most satisfactory for busy frequencies; it ensures a fair 'turn' for all ships and enables staffing requirements to be determined hour by hour. It requires, however, the close co-operation of ships' Radio Officers; in particular prolonged calling leads to delay, for a search point operator has to hold on to such calls, and while doing this he is liable to miss calls a few k/cs away. At busy periods when all working points are engaged, search point operators themselves accept weather messages and QSLs (acknowledgements of receipt) for broadcast messages. To make full use of this facility ships should, where they have such traffic, signal OBS (weather observation report) or QSL at the end of their call.

Difficulties sometimes arise in working ships fitted with crystal-controlled transmitters, as they may be requested to move up or down to a frequency differing from their crystals. For the time being, therefore, it would be helpful if such ships indicated their working frequency upon each contact. Later, when the number of such ships reaches larger numbers, it is proposed to introduce a series of indicating letters for the various working frequencies.

The greatest difficulty experienced at Portishead in expeditiously handling traffic is due to interference. A new building, now nearing completion and to be occupied on November 16th, is fitted with the most up to date technical equipment. The receivers to be used for HF are Marconi type CR150, and thirty directional aerials are to be used".

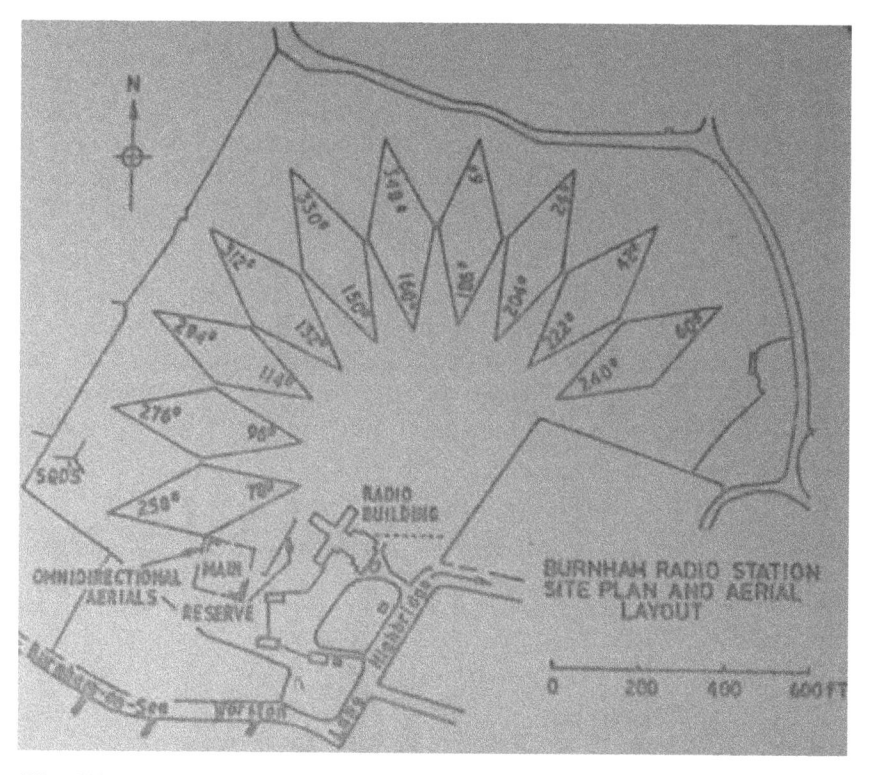

The Highbridge aerial system, 1947. SQD's are the "Stacked Quad" Aerials.

One Radio Officer remembers his first day at the station;

"When I arrived at the station in Highbridge in 1948 the operating room occupied most of the first floor of the main building. A small room on the right at the top of the stairs contained the M/F station, Burnham Radio (GRL). I had to admit that, having worked the station from 1936 until 1939, it was not quite what I had imagined it to be. The receivers were on rows of trestle tables, which were crammed together, wires scattered all over the place and the slightest nudge of a table, a cough, or other quirk of nature could cause one to lose a signal. A wire basket on the top of each receiver was the receptacle for received traffic which a messenger cleared from time to time and took to the landline room where there were two teleprinters, direct lines, one to London, the other to Bristol. I worked there for about six months before we moved into the newly built building.

The job of messenger was performed by a postman or telegram boy from the Highbridge or Burnham Post Office. Shortly after taking up duties I noticed that my basket was being emptied but could not spot anyone making the collection when I saw a hand reaching into the basket. It turned out to be a postman who was under five feet tall and hidden by the receiver!"

Radio Officer John Cave didn't find the Portishead operators so amusing, however;

"I cannot think of anything funny about GKT, the search operators always seemed so serious, while the traffic bod appeared to be harassed all the time. But it was an excellent station to work on 8 or 12 MHz, with a beautifully distinctive note to listen out for. I can also take my hat off to them after they were refurbished".

On Tuesday, November 16th, 1948 (on schedule), the old Portishead Radio receiving station at Highbridge was closed down, and the new enlarged building with new equipment was placed in operation. The new building was officially opened by the Postmaster-General of the time, the Rt. Hon. Wilfred Paling, M.P., and the event was covered by both local and national newspapers.

The GRL MF station at Burnham in the old building. Direction-finding equipment clearly visible.

An example of such coverage may be represented by the following article, published in the Western Daily Press on Wednesday 17th November 1948:

MR PALING FLICKS A SWITCH AND BURNHAM SPANS WORLD.

"The S.S. *'Capetown Castle'*, 5,000 miles on her way to South Africa, received the first message transmitted from the new Post Office radio station opened by Mr. Wilfred Paling, M.P., Postmaster- General, at Burnham-on-Sea, yesterday, and within a couple of hours had flashed back her reply. At precisely 11.50 the old radio station closed down after a ship-to-shore service lasting a quarter of a century, but simultaneously the Postmaster-General, with the touch of a switch, set into action the new station with its most up to date control and operating equipment. With the touch of the switch the indicator light winked, the station's call signs came over a loudspeaker in Morse and Post Office and Naval operators took over their new instruments. The station was 'on the air'. The opening ceremony, held in the central control room of the new 6,000 square feet pre-fab building, where guests from organisations concerned with long-distance ship-to-shore wireless, heard Mr. Paling declare: "Some of our American friends have seen the new station and have admitted they can teach us nothing about a long-distance ship-shore service". He said the ceremony symbolised a very important step in the development in both Post Office and British Maritime communication.

Mr. Paling officially opens the new Highbridge receiving station, 1948.

Burnham opened early in 1925, had installed one short-wave receiver, with one short-wave transmitter at its companion station at Portishead. In the first year traffic with ships totalled a mere few thousand words, last year traffic exceeded 6,500,000 words, and in 1949 was likely to reach the 10,000,000 mark. The station now operates a commercial service from aircraft to ground, but the volume was very light though it was expanding as time went on. He dealt with the exacting requirements of wartime, where Burnham played an exceedingly important part, and added that a good deal learnt from experience was being carried into post-war service. After the Postmaster-General had sent his message of greeting to S.S. *'Capetown Castle'*, T. A. Davies, chief of Post Office Wireless, sent a message to Capetown Radio Staff, and Capt. G. R. Waymouth, Director of Admiralty Signals, originated a message to Ceylon Radio, Later, replies were received, including one from an aircraft flying over the South Atlantic, Vice-Admiral K. S. G. Nicholson expressed, on behalf of the Admiralty, appreciation and thanks for the help received from the Postmaster-General and the Post Office in furthering the network of communications. He commented on the wonderful co-operation that existed. Mr. H. C. Van De Velde, Marconi International Marine Communications Ltd, said he did not think there existed any other industry or such close collaboration or desire to assist as in the radio marine field. Mr. J. D. Wynne, of Burnham Radio Station staff, commended to the Postmaster-General the idea of interim change of operators from ships and shore over a limited period. Facilities would be used to best advantage he said, and service would be second to none."

The PMG was compelled to write to the station manager on the 19th November;

"I was extremely pleased with all the arrangements for the ceremony at Burnham yesterday. It passed off excellently, and the greatest credit is due to everyone concerned for such an efficient and effective organisation. I will be grateful if you would convey to all of them, Post Office and Naval staff alike, my congratulations and thanks. I was much impressed by the fine spirit of all the staff with whom I came in contact at Burnham. It seems to be a very happy ship, and I am sure that the new arrangements can only serve to make it a still happier one.

Yours sincerely, W. Paling".

One local newspaper, the 'Somerset Evening World' reported extracts from Mr. Paling's speech under the headline 'Burnham's Super Radio Made Americans Gasp';

"In our anxiety to give shipping the advantages to be gained from the new technique of high frequency, or short-wave working, the Post Office installed here the first short-wave receiver. With this single receiver and one short-wave transmitter at Portishead – 24 miles away – ships suitably equipped were able to communicate direct with home regardless of their position. The going was certainly slow at first and, in the first full year, traffic with ships totalled a mere few thousand words. Last year the traffic handled exceeded six and a half million words, and the question facing us today is whether we shall reach the ten million mark in 1949. With the greater number of deep-sea British ships now fitted with short-wave equipment the prospect is certainly promising.

Already during its history and excluding its recent wartime role – which by the way is a story in itself – Burnham has to its credit many outstanding performances.

On one occasion, after a collision between the *'Doric'* and the *'Formingay'* one operator managed, I am told, to handle 208 messages in five hours without a break. Those of you with operating experience will I feel appreciate the full meaning of this achievement. I am sure that with the facilities we now have here at Burnham and with the co-operation we are getting from the overseas stations we can give a service second to none. Indeed, for once at any rate, the Americans give us the best. Some of our friends from the other side have already seen this extension and have admitted they can teach us nothing so far as long-distance ship-shore services are concerned. One of the outstanding features, as you will be able to see for yourself later, is the efficient way in which it has been possible, largely by the mutual co-operation to which I have already referred, to weld the relations and operations of the Naval and civilian staff into a harmonious whole"

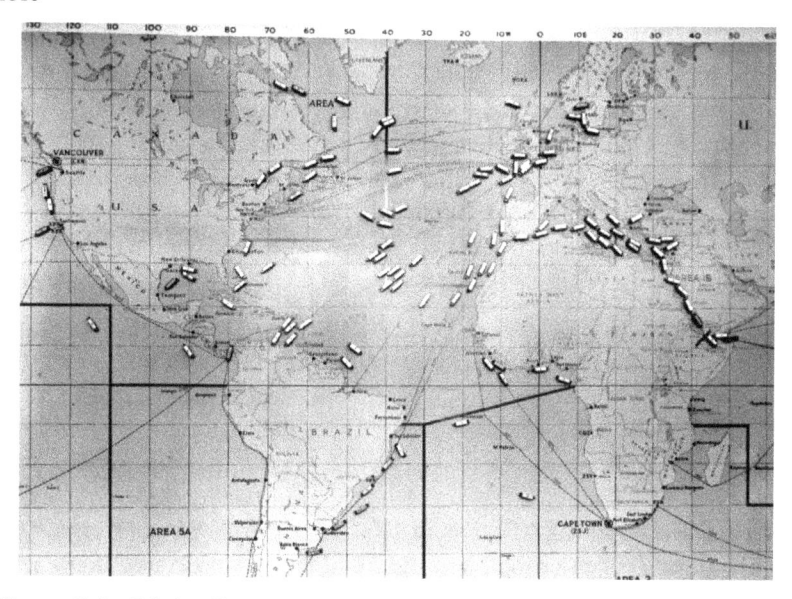

View of the Main Control Room steel map showing ships' positions and Areas.

Another newspaper also sent a reporter to Burnham, who seemed overwhelmed at the efficiency of the service in an article entitled 'The Voice of 10,000 Ships';

"In an enlarged 'pre-fab' hut down Somerset way, on a site used during the war for spotting enemy aircraft, some of Britain's best boffins are perfecting a scheme for maintaining hourly contact with any British ship sailing the world's oceans. All types of messages at the ship-to-shore radio station are recorded; a fire at sea, a changed position owing to bad weather, the birth of a son to a member of a crew. Time lag from sending the message from the station to receiving an all-clear from the ship is TWO MINUTES.

I have watched these experts at their desks at the GPO long-range station at Burnham-on-Sea. I saw ships ranging between 3,000 and 20,000 tons being 'recorded' on a wall map. One was nearing New Zealand, another off the West African Coast, another entering the Panama Canal, yet another was midway between Aden and Bombay"

Twenty-eight HF and four LF (low frequency) receiving consoles at the Highbridge receiving site were installed, and ten HF transmitters at the Portishead site were made available, together with two LF transmitters. An additional LF transmitter at Criggion was also provided. Eight of these HF transmitters were capable of providing two-channel (duplex) working. The newly installed receiving aerials now comprised of omni-directional systems for initial receipt of calls, and a 'fan' of ten rhombic and vertical-vee antennas, whose highly-directive properties made them ideal to receive signals from all parts of the world. A switching system ensured that all aerials could be used simultaneously by any number of operators without interaction, and that both omni-directional and directional aerials could be selected from any console. The aerial signals were amplified before distribution to the consoles by means of multi-band amplifiers with six narrow pass-bands corresponding to the six maritime frequency bands of 4, 6, 8, 12, 16 and 22 MHz.

The building itself comprised a main control room of some 1,296 square feet, built on to the rear of the original building. From three sides of the control room, prefabricated buildings stretched out into the station grounds. The control room, which accommodated the Ships Bureau, the circulation positions and the telephone switchboard, was a room some 36 feet square and over 16 feet in height. The room was dominated by three colour maps, covering a wall area of 976 square feet. A map of the world completely covered one wall of the room, and was flanked by maps showing the North Atlantic and the waters surrounding the British Isles. These maps showed radio receiving and transmitting stations, the main seaports and shipping routes of the world, and the chief airports and air routes. The maps were painted on steel plates, and by moving small magnets bearing the call sign of a ship, a vessel's movements were plotted.

It was a local man, George Peck, who painted the information on the maps which were initially drawn out by Naval architects. He lived on the Burnham Road in Highbridge, and his house was commandeered during WW2 to accommodate the influx of radio staff from RAF Locking.

The three wings, each having a floor space of some 1,440 square feet, housed the radio and landline equipment. The facilities comprised of 32 radio operating positions (16 in each of two wings including apparatus for high-speed working); four positions for the automatic broadcasting of traffic; 12 teleprinter positions including through-switching equipment and the various associated apparatus.

One of the large magnetic steel wall maps at the Highbridge receiving station showing the 'Areas' and the Ships Bureau in the foreground.

Each radio operating position was a self-contained unit complete with a radio receiver, transmitter cabinet, aerial selecting switches, Morse transmitting key, intercommunication facilities with other positions in the building, pencil tray, ashtray, waste-paper receptacle and even a recess underneath the desk for the operator's cup of tea.

Conveyor belts carried the messages from the operating points to the main circulation position and on to the landline room for transmission over the Inland Telegraph network. Many a Radio Officer learnt the knack of signing off the from-ship telegram (otherwise known as a 'green') and sending it on its way into the conveyor belt with one flick of the wrist; another skill lost forever.

It is interesting to review how busy the station was over the Christmas period. Up to 8 p.m. on Wednesday 21st December 1949, Burnham Radio Station had handled a total of 23,000 radio greetings telegrams, consisting of 360,000 words. Of these, approximately 20,000 came from ships.

W/T operating wing at Highbridge, 1948.

Figures from 1950 and 1951 showed the steady increase in traffic:

1950 - 444,454 messages containing 8,561,405 words

1951 - 489,270 messages containing 9,498,906 words

This showed an increase of over 10% over the 12-month period.

In March 1952 Search Points were suspended on an experimental basis which resulted in increased interference on the calling bands as ships continued to call until they were worked. However, very few complaints about long calling periods were received. A comparison of the received traffic totals during the eight weeks prior to and after the Search Point suspension revealed:

View of the new station showing the Control Room, 'A' Wing and 'B' Wing.

8 weeks ending 19th April 1952 (Search Points operating):

    52,968 messages received direct (67%)

    17,461 messages received via Areas (33%)

8 weeks ending 14th June 1952 (Search Points suspended):

    53,423 messages received direct (71%)

    15,664 messages received via Areas (29%)

The larger percentage of direct traffic since the suspension of Search Points was attributed to the increased number of working points which became available under this condition.

It is not known if this experiment was continued, although traffic levels throughout the 1950s continued to increase dramatically.

By now, a staff of 115, which included a number of Naval telegraphists, manned the receiving station, providing a 24-hour service to merchant vessels world-wide. Basically, the civilian staff were designated as RO1 (PMG 1st class Radio Officer qualification) and RO2 (PMG 2nd class Radio Officer qualification); it was the RO1's job to handle the c.w. communications, whilst the RO2s were given the responsibility of the landline and menial ships' bureau duties. However, RO2s were generously allowed to handle c.w. communications between the hours of 2200 and 0900 daily. The duty schedule at that time gave most of the night shift working to the RO2s, a situation which pleased the RO1s. Needless to say, there was a little discrimination between the RO1s and the RO2s (especially in pay differences!), but there was of course nothing to stop the RO2s from obtaining their PMG 1st class certificates whilst working at Portishead; in fact, the Post Office later granted study release time for those RO2s wishing to upgrade their certificates. The Naval operators assisted the RO1s in c.w. working. This RO1/RO2 separation continued until the early 1970s when the Radio Officer became a single grade.

Ex-Naval telegraphist Douglas (Jock) Carr recalls:

"For those of us who relished the operating part of a telegraphist's job it was a plum draft, and I was fortunate enough to be posted to Portishead Radio in 1956, a draft that was to have some consequences for my future career in the Navy.

There were, of course, no barracks, and the junior ratings were accommodated in civilian digs, paid for by the Navy, living as one of the family. These families volunteered to accommodate us, no mean commitment since watchkeeping duties meant that at times when on the overnight watch we would be asleep during the day.

Once familiarised with the receivers, the tuning controls (we could select the particular aerial(s) to give the best signal), and the administration system, we spent our time scanning our particular frequency band for ships trying to contact Portishead from the farthest corners of the world. Navy telegraphist and civilian wireless operators sent and received the signals without distinction and at was quite exciting to work the *'Queen Mary'* (call sign GBTT) or the *'Queen Elizabeth'* (call sign GBSS) whose powerful transmitters blasted their way through all interference. It was a privilege to send and receive telegrams from some of the most famous people, such as Winston Churchill, Prince Rainier of Monaco and Grace Kelly, his bride to be, when dignitaries tended to use the great liners when crossing the Atlantic. Some telegrams were in code.

Portishead Radio's principal call-sign was GKA, but a different call-sign was used for the different frequency bands, for example, (if my memory is correct) GKV (6 megacycles), GKN (8 megacycles), GKL (12 megacycles) as well as others whose particular call-sign and frequency band I cannot now recall. I used to spend a lot of my time during the day on GKV and on the higher frequencies during the night, when changes in the ionosphere demanded a higher frequency of transmission from ships in distant parts of the world. It was always a thrill to pick up a faint signal from the other side of the world... I was in my element! It was a period of doing what I loved doing without drill, divisions, mess deck cleaning, or PO's - a period of relative tranquility and calm throughout the summer months".

Ex-Station Manager Don Mulholland recalls the RO1 and RO2 situation prevalent during the 1950s and 1960s;

"I do not recall any major problems. Basically, RO2s were on a lower pay scale, as would be the case if they served on a liner. As I recall some arrangements were made in the School to help RO2s. It was common practice in the PO and in the Civil Service to have long protracted pay scales in each grade; sometimes it would take many years to rise to the top. As an example, say, an RO2 might have started at £4 a week and progressed through to say £4.75 (new money) and an RO1 would start at say £5.50. There was the incentive to progress.

When I started it was on five pounds ten shillings in January 1949. But I came in as an RO1 straight from the Post Office; probably the last of the breed. Just before the war PO recruitment for the WTS was halted, as teleprinters supplanted Morse and there was no further need for telegraphists. As I had been a Sorter Clerk and Telegraphist before the war I was able to transfer after the war, and after I had attended Hull Tech to get a PMG 1. My prior Morse work was done in the Royal Corps of Signals.

As to the division of work, I seem to recall that RO1s did broadcasts and Search Points whilst RO2s were relegated to working points. Both grades shared other work. Oh, perhaps Bureau and Control were also RO1s work but cannot be sure - so long ago and my memory isn't what it was".

Approximately eight million words of paid traffic each year were handled during this period. Daily traffic figures were recorded meticulously, giving totals of both to and from-ship telegrams, Ship Letter Telegrams (SLTs) and Naval Telegrams (ZBOs).

A Naval Radio Officer at the time, Ron Beal, recalls this period well;

"It was 1954, I was 21 years old and serving as a radio operator in the Royal Navy when I got my posting to Portishead Radio (call sign GKA).

I soon settled into my 'digs' in Burnham-on-Sea and reported for duty to GKA. This was a dream posting; I was fresh from two years spent being tossed around the Korean coast with the smell of cordite coming in the wireless office air vents, while trying to read the fastest broadcast in the world - 28 words per minute from Hong Kong (GZO).

There were 2 or 3 Naval operators in each watch of about 20+, the remainder being Post Office operators. The Naval segment was seconded to GKA originally to handle the Naval traffic from warships at sea. It is hard to believe, but the Navy had no long-range ship-shore organisation of its own; good shore-ship and world-wide base communications, but nothing to match the GKA world-wide ship-shore service.

It soon became apparent that the Naval operators would spend most of their time twiddling their thumbs if they were restricted to dealing only with Naval traffic, so it was decided that they would form part of the shift and do exactly the same work as the Post Office operators".

The new building consisted of three wings (A, B, and C), each approximately 80 ft. by 24 ft., radiating from a central control room approximately 41 ft. by 35 ft with the old buildings used for administrative and staff welfare accommodation. The central control room was the nerve centre of the station and accommodated the following three main departments:

- A 'Ships Bureau' where all to-ship traffic was routed and a comprehensive filing system containing the latest known positions of all British and foreign ships was maintained.

- A 'Traffic Control' where the daily serial numbers of traffic to and from ships were checked and information of all kinds supplied for the handling of ship-shore traffic.

- A 'Finished Check' where all traffic handled at the station was checked before being filed away.

Conveyor belts fed the traffic into this room from all receiving points in wings A and C, and to and from the teleprinters in wing B.

This room also contained three large wall maps painted on steel plates. The large central map of the world measured 14 ft. by 31 ft. and showed the divisions of the area system with the chief air and sea routes. The two smaller maps measured 12 ft. by 14 ft. and showed (i) the normal distress area around the British Isles which is covered by the short-range coast stations, and (ii) a large scale map of the North Atlantic with air and sea routes. Small magnetic holders to carry the call-sign of a ship and an arrow to indicate sailing direction were available for use with these maps which were provided mainly for air-sea rescue services.

Wing A was the main receiving room accommodating sixteen receiving positions equipped with Marconi CR150 communication receivers.

Wing B was the teleprinter and broadcast room and also accommodated in a partitioned section, the short-range coast station Burnham-on-Sea Radio (GRL).

The teleprinter arrangements comprised:

- One duplex circuit to the Central Telegraph Office, London.

- Nine through automatic switching circuits (three in reserve). This system enabled the operator at Portishead to set up his own connection, and transmit the message directly to the terminal teleprinter office, dispensing with the assistance of intermediate offices.
- Three circuits to Admiralty London.
- One circuit to Dunstable Meteorological Office.
- One circuit to Lloyd's of London.

Four positions in this wing were available for the broadcast transmission of traffic to Areas 1A, 1B and 1C.

Wing C was the reserve receiving room accommodating sixteen receiving positions, twelve equipped with Marconi CR150 communication receivers and four with Marconi CR100 receivers for LF working (GKU). Four HF and two LF receiving positions were equipped with high-speed recording apparatus.

A Naval posting or 'draft' to Burnham was regarded as a 'good number' under the RN New Centralised Drafting system, appealing particularly to natives of the area.

Except for Senior Courses, few Communicators got the opportunity to look around Burnham ship/shore wireless station whose dual role combined that of the principal station in the World-Wide Merchant ship long-distance organisation and as the Admiralty Ship/Shore receiving terminal.

The station, which is under GPO control, was staffed mainly by Post Office personnel. There were 22 Naval operators which constituted no more than 25% of the operators. A 4 Watch watchkeeping system was employed, Naval telegraphists working alongside and carrying out the same wireless watch duties as their civilian counterparts. Naval personnel did not exclusively work Naval Ships but spent the greater part of their time on duty working commercial ships whenever practicable. However, Naval Ship working was always carried out by Naval telegraphists.

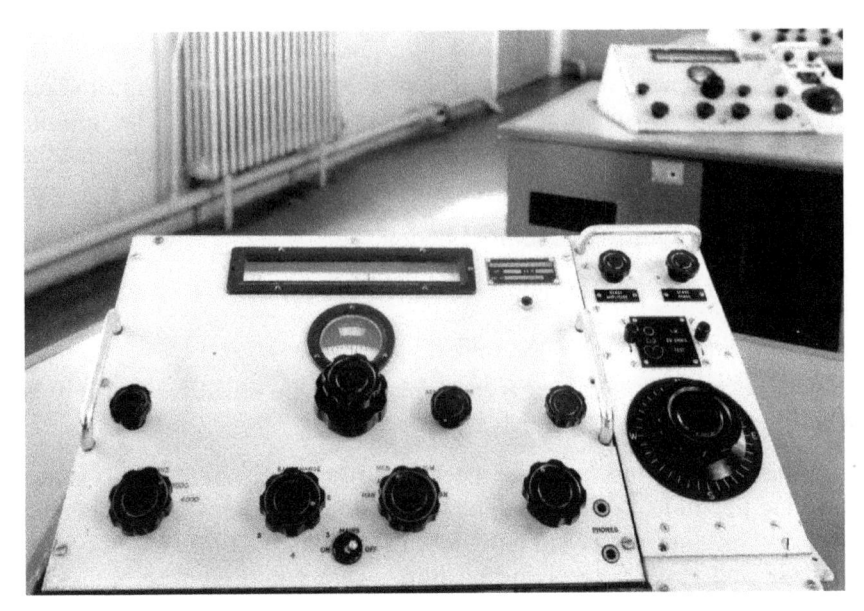

A Marconi CR100/2 receiver installed on a W/T console with the aerial selection equipment to the right.

It follows that a pre-requisite required from telegraphists drafted to Burnham was a good standard of operating. Ships of all nationalities and denominations would be worked and handling messages as part of the public service called for a high degree of competence and operating ability.

There were 32 operating positions at Burnham for dealing with ships' incoming wireless traffic; of these, 28 were for working long distance on the higher frequencies, the remaining 4 for low frequency working. In addition, a coastal station service was provided for short-range shipping on 500 kHz, call sign GRL.

The methods of searching for ships in the new station varied from that detailed earlier in this chapter. The rapid increase in the number of ships equipped with HF apparatus in the post-war period created conditions that could only be met by an extension of facilities; Search Points were maintained in four of the frequency bands simultaneously during certain periods of the day.

The main advantages of the system were considered to be;

(a) The coast station had a ready means of ascertaining immediately the loading in the various bands and could allocate staff accordingly.

(b) If a ship was answered promptly by a coast station, a certain amount of anxiety on the part of the less experienced Radio Officer was dispelled and he would 'queue' with reasonable patience, even if allocated a high QRY (turn).

(c) The amount of signalling in the calling band was very much reduced and consequently the amount of interference.

(d) The maximum number of ships was answered, irrespective of signal strength or nationality and without discrimination.

(e) The sequence in which ships are worked was systematic and generally, the working is more orderly.

Portishead Radio therefore extended the periods of Search Point operation in the four main receiving bands to the following schedule:

    16 MHz Band   0800 - 1800 GMT

    12 MHz Band   0800 - 2000 GMT

    8 MHz Band   0800 - 0100 GMT

    6 MHz Band   0800 - 0100 GMT

Marconi CR150 communication receivers were used at the six search and twenty-two working point positions; each position was numbered. The receivers were sunk into tables with the receiver control panel at an angle of thirty degrees to the horizontal and located suitably for left-hand operation. A switch panel for selecting aerials and test signals was located adjacent to the right-hand side of the receiver. The Morse key, microphone for intercommunication, transmitter control unit and Search Point control unit giving additional intercommunication facilities, were located to the right-hand side of the position.

The aerial selection unit provided switching facilities to any of the aerials and also accommodated the switch for the test signal oscillator. Selection of either the omni-directional or directional aerials was made by the operation of a lever-type switch; when directional aerials were selected, a rotary switch marked with the cardinal points of the compass was then used to select a particular direction.

Crystal oscillators provided continuous dot markers at the centre of each of the ships' calling bands and could switched at will by operating the 'Test' lever on the aerial selection unit.

The transmitter control unit provided facilities by means of lever type keys for the immediate control of any transmitter. A red monitoring lamp on the send/receive switch followed the keying and indicated when a transmitter was engaged.

The Search Point control unit provided facilities via the intercommunication system, for calling and speaking to all other search and working points in the same wing and to the control room. Calls from working points were indicated on white calling lamps associated with numbered keys to correspond with the receiving positions. Signals from working points to indicate that a working point is disengaged and is listening on the intercommunication line for the next ship were shown by green lamps. Also, output signals from the Search Point receiver would be extended to a working point when the intercommunication connection was already established by depressing a plunger on the control unit.

In the control room, an operator checked the serial numbers of all incoming and outgoing messages, distributed the traffic to and from the teleprinter room, and answered all queries from search and working points over the intercommunication system.

The actual operation of a Search Point differed little from that used prior to 1948. Each Search Point was allocated a number of working points dependent on the number of ships waiting to be worked in a particular frequency band. Because the Search Points and the associated working point normally shared a single transmitter it was generally felt that no more than five working points should be allocated to any one Search Point unless an additional transmitter became available; this would prevent the inevitable chaos of numerous ships being worked on one transmitter.

The Search Point operator maintained a log sheet on which was recorded details of each call received as follows;

- Time of call.
- Call sign of ship.
- Direction aerial.
- Working frequency.
- QRY (turn number)
- Working Point number allotted.
- Time passed to working point.
- Remarks.

The Search Point operator searched over the calling section on his particular frequency band using an omni-directional aerial. Upon intercepting a call, he would have switched to a horizontal rhombic (and highly-directional) aerial to ascertain the optimum signal strength of the calling station. Upon receipt of the vessel's working frequency, he would then advise the vessel his turn in the queue (QRY) and pass the relevant details via the intercommunication system to the next available working point.

A Marconi Mariner article of the time gave a stern lesson in how to call;

"Provided that the calling station has indicated its working frequency, the appropriate 'QRY' is signalled and when this has been acknowledged by the ship the search is continued. It will now be obvious that unnecessarily long calls and failure to signal the working frequency at the end of the call will considerably slow down the speed of working of the Search Point and add to congestion and interference in the calling bands, particularly during the busy periods of the day"

An experienced Search Point operator could handle as many as forty calls in one hour provided he received the co-operation of the ships as quoted above.

When a working point became free, he informed the Search Point operator via the intercommunication system. He would then be passed details of the next ship on the list; call sign, optimum directional aerial bearing, and working frequency. Once received, the working point operator would then request the ship to 'QSV' (send a series of V's) on the working frequency until communication was established, ensuring that the ship's call sign was quoted indispersed within the characters.

It will now be appreciated that the Portishead operator who eventually took the traffic was not the operator who answered the original call and signalled the 'QRY'.

Any traffic on hand for the vessel would be passed on request to the working point from the control room, again by use of the intercom, for delivery to the ship using a dedicated frequency. All traffic from the ship would be transcribed directly onto a telegram form by typewriter, and upon receipt (QSL), passed via the conveyor belt to the teleprinter room in wing B for despatch via landline to the Central Telegraph Office (CTO) in London, or through a company's private wire link. Lloyds of London also had its own dedicated teleprinter link. Weather reports were sent to the weather centre in Dunstable along their own dedicated link. Vessels' position reports (TRs) and notification of change of area were also passed down the conveyor belt to the Ships' Bureau position for updating.

Commercial working with ships was mainly unremarkable; stores orders, ETAs, fuel requirements, etc. were the content of the majority of telegrams. However, there were some lighter moments, as Ron Beal recalls;

"We might work in turn a warship with one small message, and then a liner with 60 telegrams! I remember it also took some skill to master the fact that 6 operators were keying the same transmitter - the only help being a little red light alongside the key - this was duplex, remember, and one never heard one's own Morse (no sidetone facility).

I had no experience of commercial working and it took some time to get used to strange procedures and even stranger Morse code. Two incidents in the same shift that I shall never forget will explain what I mean.

The Search Point op gave me the next call sign on his list - 'KJEH'. I moved him to his working frequency and gave him a 'K' (invitation to transmit) to start sending the first telegram. It was our practice to type the telegram straight onto the regulation form. The first block on the form measured 3" x 1/2" and this was for the ship's name. I received the letters 'US' for her name and the rest of the message was OK. I thought I had misread the name (as happens) and gave the usual 'QRA? - 'What is your ship's name?'

After a few seconds silence she replied (I am trying to fit this into the empty box) "This is the United States Steamship *'United States'*. Blue Riband holder for the fastest Atlantic crossing and the pride of the Merchant Fleet".

I had dared to ask the mighty U.S. who she was! I apologised and was forgiven.

A little while later - still smarting from my encounter with 'KJEH' - I was given 'HPLU' to work. Having shifted her to her working frequency I gave her the 'K' to pass her telegram; a few seconds later she was drowned in giant QRM (interference). I attempted to break in and stop her but she just ignored me and finished her telegram. The QRM disappeared just in time to get the signature. I was annoyed and complained she didn't listen for break-in. The op then astonished me with the remark "if you can't read Vibroplex please say so". The steam came out of my ears as I replied: "I can read anything you can send" (wasn't I the hero of the fastest broadcast in the world), then invited him to pass his message. This guy then sent a 40-word telegram in seconds, in perfect code that was nearly a blur. For the second time that day I apologised, he repeated his telegram, and we became friends - so much so that Manuel de Leon of the S.S. *'Poukoulet'* (HPLU) asked for me by name every time he called GKA - it was very embarrassing".

Many of the operators at Portishead would recognise the call signs being used, and strike up a friendly relationship with the Radio Officers on board. Radio Officer Mike Quan recalls an incident from the mid-1950s;

"When I was on the cable ship *'Norseman/GBVS'* one of the R/Os at Burnham asked my name. He had been on the same ship before me. When he found out I lived in Weston he used to give me free passage of notes to my mother, which he used to call over to her on the 'phone and then get a reply. From time to time I used to come up in the traffic list as 'GBVS PSE QSO direct' which meant there was a note for me (this was an instruction for the vessel to contact Portishead direct and not via one of the area stations).

Eventually, when I left the ship at Gibraltar, I had to wait for about a week for a passage home (air travel wasn't what it is now). At only a few hours' notice, I was told to join the 'Otranto' which was then a troopship laden with Palestine police and RAF personnel coming home from Port Said. When we sailed, I made a trip to the radio room to send a telegram home; but there was no need – when they saw the name they said 'GKL has already asked if you were on the passenger list and we said yes – and gave our ETA Southampton'. What a service!!"

Mike goes on to relate;

"At one time I had the magic call sign GDSS – not only was it a beaut to rip off on the key, but the *'Queen Elizabeth'* was GBSS. I can't remember how many times when I called GKL (Portishead), WSL (Slidell) or WCC (Chatham), they came back with GBSS! I certainly got attention in the crowded bands of those days".

Another seagoing Radio Officer, G.A. Fido, remembers the trauma of working Portishead Radio in his early days;

"I was a seagoing R/O from 1956 to 1968, with a break of 6 years in between. The area scheme was in force in the earlier part of my career, and Portishead Radio, to an 18-year-old, was omnipotent. The older hands used to impress upon me the necessity for correct procedures. Failure to do so could mean that Portishead would report me and I might end up losing my ticket.

Imagine then my horror in 1957, on my first deep-sea trip with Blue Funnel, as we passed down the Irish Sea and came within range of Burnham Radio, I was requested to contact Portishead at the earliest possible opportunity.

Had they found out about my last ship's captain giving a position 200 miles south of Land's End so that he could ring his wife on the radiotelephone, when in fact we were just out of Las Palmas?

Had they queried my 'nothing heard' log entries during a quiet crossing of the Pacific when the only other ship for three weeks was the *'Captain Cook'*, and I had left it a bit late to change areas?

However – I eventually bit the bullet, and contacted Portishead as requested, only to be directed to an operator who was trying to track down one of his friends, who when he had last heard of him, had been with Blue Funnel".

A word here about the aerials in use at the time; for the most efficient method of operation, the Search Point operator would use an omni-directional receiving aerial facility, but the working point operator would need directional reception. An ingenious aerial distribution and amplification system was therefore utilised, and remained in use (with some variations and additions) until the new operations centre was opened in 1983.

The aerials (located at the rear of the receiving centre at Highbridge) can best be described from an information sheet available during the mid-1950s;

"Aerial facilities are provided at Portishead Radio by the following three groups;

(a) Six omni-directional horizontal V-dipoles of multiwire cylindrical cage construction, sufficiently aperiodic to enable one aerial of each type to cover one frequency band. Six vertical dipoles are also available in reserve.

(b) Ten bi-directional horizontal rhombic aerials, giving in effect twenty directional aerials, each serving two diametrically opposed zones 18 degrees wide, and providing all-round coverage of 360 degrees. These aerials are sensitive to signals of normal downcoming angles.

Bi-directional reception from these ten horizontal rhombics is achieved by connecting both ends of each aerial to the receivers via an aerial amplifier and distributing system; the impedance of the aerial amplifier, at marine band frequencies, is such that it suitably terminates the aerial and replaces the normal absorbing resistor termination.

The ten aerials are erected at a height of 102 feet wide, side length 246 feet, and side angle 140 degrees. They are arranged fanwise round the station, thus effecting economy in site area and supporting structures, and ensuring a reasonable degree of freedom from mutual interference between aerials.

(c) Five bi-directional half-rhombic vertical aerials (inverted V's), serving two diametrically opposed zones 36 degrees wide, providing all-round coverage of 360 degrees. These aerials are interspersed with the horizontal rhombic aerials and are especially sensitive to signals of low downcoming angles, particularly at the lower frequencies, where the response of the horizontal aerials at very low angles is poor.

Bi-directional reception is provided in a similar manner to that used for the horizontal rhombic aerials.

Aerials (b) and (c) are brought into the station on transmission lines of 600 ohms impedance. They are transformed to 75 ohms at a gantry and fed to aerial amplifiers in wings A and C. Each amplifier has six narrow pass-bands covering the ships 4, 6, 8, 12, 16, and 22 MHz frequency bands, each band being separately amplified.

Signals from the multi-band amplifiers are distributed at an equal level and without interaction, to all H/F receiving positions via a transformer and resistance network, and are separately connected to the aerial selection unit.

The distribution from the omni-directional aerials differs slightly from that of the rhombic aerials, in that signals pass through a combining unit before being fed to the amplifier and then to the aerial units in one cable.

Every H/F receiving position has immediate access to all of these aerials via its associated aerial selection unit and any number of operators may use the same aerial without interaction".

A marked increase in traffic occurred in 1956 during the Suez crisis, when British Naval and merchant ships used the services of Portishead Radio to send telegrams home to their families, an early example of how Portishead's facilities could be used to boost morale in times of conflict. Inspection of the daily traffic figures for this period indicated the increase in traffic; over 1,000 from-ship telegrams were taken on two consecutive days in early August, compared with the normal daily average of approximately 700. To-ship telegrams also increased in number, reaching over 600 items per day, compared with the normal average of around 450. It should be stressed, however, that these figures include items which came to Portishead over the Admiralty lines as part of the 'Area' scheme described earlier in this chapter.

Portishead R/O Phil Mitchell remembers one distress incident which occurred around this time;

"In or around 1954, there was a near distress situation involving the *'Moreton Bay'* and a sister ship operating in the Indian Ocean off Mauritius. In the early hours of the morning GMT, GKT intercepted calls from one of the passenger ships indicating she had a fire onboard and needed assistance. We initiated calls to the sister ship and managed to divert her to assist. The owners were, needless to say, very gratified that we had kept the situation 'in-house' and GKT got a very nice letter from them to that effect. The R/O who intercepted the calls at around 0500 jumped his turn on the 'fiddle sheet' and was sent home early in recognition of the deed".

By the mid-1950s the service was showing an increasing deficit due to rising costs, and the decision was taken to increase the tariff rates considerably. These rates, which came into effect on 1st January 1956 were;

Full rate: 1/6d per word

Reduced rate: 9d per word

SLT: 5d per word (minimum charge 8/4d for twenty words)

A medical advice (MEDICO) service had been provided at Portishead for some time. This facility ensured that vessels who were in urgent need of medical advice were either connected by radiotelephone to the duty doctor at Weston-super-Mare hospital, or a telegram was sent to Portishead which was telephoned directly to them. The Doctor's reply was then sent back to the vessel and any questions replied to. No charges were raised for 'MEDICO' calls or messages, and this service remained free of charge throughout the life of the station.

The Suez crisis in 1956 brought a greatly increased level of traffic to Portishead, as Mr. C.H. Stanley notes in a letter;

"At the time of the skirmish between Britain/France and Colonel Nasser of Egypt, a number of British ships were trapped in the Suez Canal lakes. Their only contact with the UK was via Portishead Radio; dates, times and frequencies escape me now, but I listened regularly and know how important those crews found Portishead to get news to and from their folks".

1957 saw the end of the LF (low frequency) service; use of these frequencies had dropped off considerably since the end of the 1940s, as the majority of vessels were now equipped with HF apparatus. The transmitters were subsequently converted to HF use to cope with the ever-increasing volume of traffic. This became noticeable at sea, as R/O B. Samuel recalls;

"When I first went to sea in 1956, I remember the Portishead transmitters had a very distinctive 'growl' which made it easy to pick out from other transmitting stations. Thereafter the note changed and became the same as other stations, and I always considered this a retrograde step".

The Portishead transmitting station also took delivery of 4 STC 5 kW DS12 transmitters at this time.

It was also discovered that owing to a costing error, the amount charged to the Ministry of Transport for safety of life services had been overstated. To cover the deficit resulting from rectification of this error, the full rate telegram charge was increased to 1/6d per word with effect from 1st October 1957. No changes were made to the reduced rate or SLT tariff.

Meanwhile, the medium-range service was going through extensive change. Radiotelephone equipment was being installed at all the U.K.'s coast radio stations, and new stations at Oban (1949), Stonehaven (1958) and Ilfracombe (1959). The latter station had in fact been located in a temporary home in the attic of the Ilfracombe Head Post office since 1955). It was also in 1959 that the medium range service at Portishead (Burnham-on-Sea Radio) was closed down.

To give an idea of how quickly the medium range radiotelephone service expanded, a list of 'start of service' dates will be of interest;

Humber Radio (GKZ)	February 1937
Portpatrick Radio (GPK)	September 1937
Cullercoats Radio (GCC)	1st October 1946
North Foreland Radio (GNF)	1st November 1946
Niton Radio (GNI)	5th February 1947
Land's End Radio (GLD)	1st July 1947
Wick Radio (GKR)	17th September 1947
Stonehaven Radio (GND)	1st November 1947
Oban Radio (GNE)	1st October 1949
Ilfracombe Radio (GIL)	1st November 1955

This service was primarily intended for small craft on short voyages, but calls were accepted from passengers on short voyage passenger ships in addition to those on ship's business. Calls were also accepted on ship's business and with ship's crews on larger ships on long voyages. Over 100,000 calls were handled in the medium range service between ships and telephone subscribers in 1958, which proves the value of this service.

The VHF (short-range) radiotelephone service was introduced early in 1957 whereby suitably fitted ships could call directly to the telephone exchange in Gourock and be connected to telephone subscribers on shore. This was introduced when it became clear that the service offered by Portpatrick Radio was found to be inefficient when vessels were located in the Clyde estuary.

On 5th September 1958, a VHF service was introduced at North Foreland Radio, and this type of equipment was subsequently installed at other coast stations between 1959 and 1967.

December 5th 1958 saw one of Portishead Radio's most historic days; Her Majesty the Queen, accompanied by The Postmaster-General Ernest Marples, visited the station.

On the occasion of the visit by
HER MAJESTY THE QUEEN,
to Burnham Radio Station
on Friday, December 5th, 1958,

THE RIGHT HONOURABLE ERNEST MARPLES, M.P.
Her Majesty's Postmaster General
requests the pleasure of the company of

R.S.V.P. to
The Inspector of Wireless Telegraphy,
G.P.O. Headquarters,
St. Martin's-le-Grand,
London, E.C.1.

9-50 a.m.
for 10-10 a.m.

Lounge Suits.

Official invitation to the Queen's visit to the Radio Station, 5th December 1958.

Looking at the paperwork involved in the visit, it would seem that nothing was left to chance, apart that is from the 'Radio Officer' factor. Numerous local dignitaries were invited, many of whom had no involvement with the station which caused much discontentment amongst the staff. The following letter from Ken Wilson, local secretary of the UPW, was sent on 22nd November 1958;

"To the Officer-in-Charge, Burnham Radio.

I am instructed to write to you on the subject of the visit of H.M. the Queen and H.R.H. the Duke of Edinburgh. My committee feel that the list of presentations include persons only remotely connected with this station, to the exclusion of officers who have loyally served this department for many years.

The committee are disgusted that the name of our Assistant Superintendent does not appear on the list, and no member of staff. We would ask you, even at this late stage, to press for the inclusion of Mr. W.G. Halford and at least one staff member. I must inform you that feeling on the station is running very high and we ask that immediate representations be made to meet the staff wishes.

If our wishes cannot be met, my committee feel that they will be justified in making a direct appeal to the Postmaster-General.

Yours sincerely,
K. Wilson, Local Secretary".

The local engineers also jumped on the bandwagon, and on 24th November, a letter was sent from their Branch Secretary to the Burnham Engineer-in-Charge;

"In connection with the Royal Visit on December 5th, I have been asked by the POEU members to inform you that there is a strong feeling of resentment against the list of people proposed for presentation. We feel most strongly that it is completely wrong for people who have only remote connections with Burnham Radio station to be presented, whilst others, who have given loyal service for a number of years are not included.

It appears to us that the local staff of the Engineering Dept. are being, once again, completely ignored, and I appeal to you most earnestly to do all in your power to get this oversight rectified in time.

Yours faithfully,

W.A.L. Forsdike (Branch Secretary)".

The management reply on 26th November was typically short;

"The DG has seen these papers. He decided:-

(a) A member of the operating staff should be presented while the Queen tours the station;

(b) No engineering officer should be presented from the rank and file staff (Either Mr. Anderson or Mr. Wadsworth (Engineers-in-charge) is still to be presented of course)

(c) I have since been told that Mr. J.A.J. Blatchford is the operator to be presented".

Thankfully there were to be no further complications and the royal visit went smoothly and on schedule.

Radio Officer Reg Hawkins being observed by HM the Queen sending the telegram to the *'Empress of Britain'*.

A telegram was sent on behalf of Her Majesty to the vessel *'Empress of Britain/GVCN'*, which read:

CAPTAIN J.P. DOBSON

MASTER

S.S. EMPRESS OF BRITAIN =

FROM BURNHAM, WHERE POST OFFICE RADIO SERVICES MAINTAIN COMMUNICATION WITH SHIPS ALL OVER THE WORLD, I SEND MY BEST WISHES TO THE EMPRESS OF BRITAIN AND ALL WHO SAIL IN HER ON THIS HER 43RD VOYAGE SINCE I LAUNCHED HER THREE YEARS AGO

= ELIZABETH R

The vessel replied immediately with the following;

HER MAJESTY QUEEN ELIZABETH II =

THE SHIP'S COMPANY OF THE EMPRESS OF BRITAIN RESPECTFULLY OFFER TO YOUR MAJESTY THEIR HUMBLE DUTY AND ARE HONOURED BY YOUR MAJESTY'S GOOD WISHES AND CONTINUED INTEREST IN THEIR WELL BEING.

= DOBSON, COMMODORE R.N.R, MASTER

Of course, this was a carefully stage-managed event, although local management were quick to reclaim all expenses. The official visitors' book was duly signed by members of the Royal Party (including Her Majesty), and there was excellent local and national media coverage, with reports appearing in numerous newspapers at the time.

Derek Osborn, a recently-recruited R/O at the time, recalls the visit with mixed emotions:

"When it was first planned, it was agreed that a telegram would be sent to the *'Empress of Britain'* by the operator who happened to be rostered on (I think) Point 8 at the time of HM's arrival. That person was me!

As I had only been at GKL for 9 months I thought this was extremely unlikely. And so it came to pass. Len Froud (O/C at the time) called me into his office. He said he thought perhaps a more senior person should have the honour, so Reg Hawkins (Jim's father) was appointed. As a consolation, I sat on the point in front.

The Duke of Edinburgh failed to turn up. He was fog-bound in Bristol. If my memory serves me right the Queen went on to Bristol to inaugurate Standard Trunk Dialling (STD). Prior to that, all calls were connected by a telephone operator.

Staff families congregated in the drive leading to the building and were given flags to wave."

It is interesting to note the costs of making radiotelephone calls at the end of the 1950s. Calls from ships in Zone 1 (ships located within a 1,000-mile radius of the UK, including the Mediterranean) were charged at £1 16s 0d (£1.80p) for a 3-minute call, and ships in Zone 2 (the rest of the world), £3 0s 0d exactly, again for a 3-minute call.

Her Majesty Queen Elizabeth II signs the Portishead Radio visitor's book.

High-frequency radiotelephone services were not handled at Portishead, but at the Post Office radio station at Baldock; these services did not come to the operational centre at Highbridge until 1970.

The Ship Letter Telegram (SLT) service was very popular at the time. This was, in effect, a cheap way of sending private messages to friends and family, although some ships did use the service for non-urgent ship's business. Upon receipt at Portishead Radio, the message was sent to the destination by first class post, therefore saving the inland telegraph charge.

Press radiotelegrams were also common; remember that this was way before the era of satellite communications, and press reports to and from ships had to be sent as radiotelegrams. The rate (again in 1959) for these messages was 9d per word for 'deep-sea' ships and 4½d per word for 'short voyage' ships and fishing craft.

The 50th anniversary of the service was commemorated by the publication of a souvenir booklet by the Post Office, copies of which were made available to all staff.

The 50th Anniversary booklet published by the Post Office in 1959.

Important changes in landline communications took place towards the end of the 1950s; previously, messages to ships had passed via local Post Offices to the various coastal stations, while other countries were accessible through the London Central Telegraph Office (CTO). Portishead Radio introduced teleprinters working directly into the CTO, making it possible to speed up communications between the ship at sea and the sender/receiver on shore. The telex system was improved so that private subscribers were able to tender their messages directly to Portishead Radio for onward delivery to the ship by radiotelegram. Some companies utilised private wire circuits into Portishead to handle the high volume of traffic being sent and received to and from their ships, a prime example being the Meteorological Office, who used Portishead for weather routeing, observations, and bulletins to ships.

The decade ended with exceptional to-ship traffic figures of 269,000 ships contacted with 516,000 messages sent. Further improvements were still on the horizon, however, and the 1960s would prove to be an exciting decade for the station.

CHAPTER FOUR

THE 1960S

The early 1960s saw radical changes and the introduction of new facilities to the Post Office Maritime Radio service. The medium range radiotelephone service was extended to passengers on long voyage ships through Land's End Radio (and subsequently Niton Radio) on 1st February 1960, and a 'unified' Post Office Ship's Telephone Service (incorporating the short range (VHF), medium range (MF) and long range (HF)) was introduced later that year.

"The national and local press gave details of this new radiotelephone service and also mentioned the long-distance radiotelegraphy service, as the following extract from February 1960 illustrates:

"A new telephone service between passengers on ships at sea and telephone subscribers in the United Kingdom has been introduced by the Post Office. It has been made possible by equipping Land's End radio station with additional radio channels. Niton Radio will he similarly equipped in April, and North Foreland Radio and Anglesey Radio will follow. With a nucleus of eight stations bought from the Marconi Company and Lloyds, the Post Office started its marine radio communication system in 1909. In those early days, there were only 286 British ships equipped and the small staff of the stations dealt with about 50.000 messages a year. Today, 850,000 messages are handled and act as the shore link with 6,300 British and 3,000 foreign ships.

Although some of the stations have been re-sited, most of the original tenes are still working and three new stations are probably the most up to date in the world. From the International Telephone Exchange in London, a long range Ships' Telephone Service is provided for ships anywhere on the seven seas. In addition, Burnham Radio operates an area scheme for long-distance radiotelegraph work to Commonwealth ships. The system, developed during the war, enables the station to operate, in effect, as a pageboy of the air to transmit at regular intervals a list of ships for which messages are waiting. One of the really important uses of the Coast Radio Services is that which is given free to ships of all nations. It is the distress service which comes to the aid of ships in peril, or when injury or illness make expert help essential. The constant alert kept at the stations round-the-clock enables immediate help to be rendered and, during a distress call, normal working is suspended so that quick assistance can be given."

A system of zones was introduced;

On 1st May 1960, the charging of calls was modified to integrate the service, the only variations being in the long range service when ships were outside of the range of the medium range stations.

- **Zone A** - all northern waters within the limits 12 degrees north, 7 degrees east, and 46 degrees north as the southerly limit and no northerly limit.
- **Zone B** - all northern waters beyond zone A and within the limits 35 degrees west, 35 degrees east, 35 degrees north and the whole of the Mediterranean Sea.
- **Zone C** - all waters beyond those defined in zones A and B.

It should be remembered that the long range HF radiotelephony service at the time was not operated at Portishead Radio, but from the GPO receiving station at Baldock, using transmitters at Rugby and Criggion (which was the back-up transmitter location for Rugby); it was not until 1970 that the HF R/T service transferred to the Portishead Radio control centre at Highbridge. Vessels requiring radiotelephone calls would send a service message (ATEL) via Portishead Radio at least 24 hours in advance, who would then relay the message to Baldock with the relevant information to set up the call.

Staffing at Portishead was still made up of ex-Merchant Navy Radio Officers and also those from the Royal Navy, as Bob Woods recalls his first posting;

"Highbridge in 1960 was considerably different from 'The Last Picture Show' image it presents today.

The railway station still had several platforms (it had once had 8 or 9 believe it or not!). There was the up and down line that exists to this day, but there was also the branch line to Burnham and another that went to Bason Bridge. The station had waiting rooms, a booking hall and a staff, most of whom spent a good deal of their working time in the nearby Cooper's Arms! The Docks were still open to a limited extent. The railway workshops were still open as was the Bacon factory. A number of pubs, now closed, were open and for the latest Ealing comedy there was the Regal cinema! It was close to the cinema that the railway level crossing (with its gates) crossed the road on the way to Burnham where the station was more or less opposite to the Somerset & Dorset pub. The track for most of its length followed roughly where Marine Drive now runs. In the summertime the train was full of day-trippers and children waving buckets and spades. It was strictly a 'Thomas the Tank Engine' school of rail travel! Mums and Dads were dressed in what appeared to be Sunday clothes, more suitable for a wedding than a day on the beach. There was little in the way of trendy, 'designer' clothes 40 odd years ago!

When our sailor boy arrived at the railway station at Highbridge he might well have been given the telephone number of a Mrs. Herring. She ran the local taxi service - the only taxi service I should emphasise! The Herrings also ran the local 'luxury' coach service. The office and garage were in Adam Street Mrs. Herring was also a big cheese on Burnham UDC (Urban District Council). Burnham set its own rates in those distant days! She was later to become Mayor and all sorts of local affairs featured on her agenda. This is not to say she was bad - just prominent locally.

Without checking I would guess the extent of the local population in the early 1960s would have been five to ten thousand against more than double that today.

Anyway, once the taxi had eventually arrived our sailor was transported to the station and digs would have been arranged or a list of boarding houses would be given to him to make his own arrangements.

I was given an address in the Burnham Road and during the following year or so lodged at two or three other houses in the town. The sailors, most of whom were single, were a good source of regular income for local landladies. Sailors then had to wear uniform when on duty and were a common sight around the town - this had some advantage as we could easily be spotted by the local girls who were generally only too happy to get to know us. In the early sixties there was no Pub culture for young people as there is today. One got to know each other by a stroll along the sea front or the week's big event - the Saturday dance at the Winter Gardens at Weston-Super-Mare. (I once met and got to know Susan Maugham (of Bobby's Girl fame)) who at the time was a resident band singer there.

There was a good social scene for the sailors and it was not unusual to emerge from the station coming off duty to find one or two girls hanging around the front gate...Ah!! Those were the days!

But many of the sailors (including myself) found themselves girlfriends and eventually a wife from the district.

Popular local pubs for the sailors were the Commercial Hotel and the 'Ring of Bells' in Oxford Street, now long demolished but stood where the large LIDL sign is now erected. Skittles teams were joined and acquaintance made with the local cider (at that time cider, unlike beer, carried none or very little tax) so was it was very cheap, popular and strong!

So to conclude, Burnham Radio was a very nice draft chit for most of us. Plenty of home comforts that were a long way from life on the forward mess deck of a destroyer where sleeping and recreational spaces had to fit in around machinery. Today of course, and rightly so I might add, warships are designed with crew accommodation in mind from the start".

Despite increased traffic levels, all telegram rates underwent a revision process in 1960, and from 1st September that year, the rates were fixed at;

STANDARD RATE:	1/8d per word (7d ship charge, 10d coast station charge, 3d inland telegraph charge).
REDUCED RATE:	1/- per word (3½d ship charge, 5½d coast station charge, 3d inland telegraph charge).
STANDARD PRESS:	10d per word (3½d ship charge, 5d coast station charge, 1½d inland telegraph charge).
REDUCED PRESS:	6d per word (1¾d ship charge, 2¾ coast station charge, 1½d inland telegraph charge).
SLT:	10/- per 20-word message (4/2d ship charge, 5/10d coast station charge). Each extra word charged at 6d (2½d ship, 3½d coast station).

Later that year, on 1st October, the medium and short-range services were extended to allow calls between ships in Zone A and the continent of Europe and subsequently with subscribers anywhere in the world; remember that previously calls were only available to subscribers in the United Kingdom.

The radiotelegraphy service continued to expand rapidly during the early 1960s. A glance at the traffic figures at that time makes for most interesting reading, for example;

Week commencing 3rd April 1960

5,381 from-ship telegrams

2,444 to-ship telegrams

2,879 Ship Letter Telegrams (SLTs)

131 Naval Telegrams (ZBOs)

Week commencing 18 December 1960

7,882 from-ship telegrams

11,402 to-ship telegrams

5,546 Ship Letter Telegrams (SLTs)

50 Naval Telegrams (ZBOs)

Don't forget, this was Christmas week, a traditionally busy time for telegrams. Even so, the handling of over 22,000 telegrams in a one-week period was quite phenomenal.

The corresponding week of 1961 featured an even higher total of telegrams handled, this time over 24,000, including 8,539 from-ship telegrams. As one can imagine, it was not uncommon for a build-up of ships on each band to exceed over 50 (with an associated delay of well over 2 hours), and Search Points were often closed to enable the operators to catch up on the delay.

This pattern continued throughout the 1960s, with traffic figures averaging between 13,000 and 17,000 items per week, peaking at over 20,000 during the Christmas period.

The area scheme referred to earlier in this book continued to operate successfully. The GPO 'Handbook for Radio Operators' from 1965 referred to the scheme as the 'Long Range Area Communications Scheme', and gave comprehensive details of the service.

The publication stated that:

"The scheme provides for a long-range ship/shore radiotelegraph radiocommunication service in which Commonwealth countries, the Irish Republic and the Republic of South Africa take part. Foreign ships are not normally permitted to participate, but foreign ships on British Government charter may do so provided that prior application by the appropriate Government Department has been approved by the Radio Services Department.

Each area is served by an area transmitting station and one or more area receiving stations as shown in the following table:

AREA	TRANSMITTING STATION	RECEIVING STATION	SUPPLEMENTARY RECEIVING STATION
1 (1A, 1B, 1C)	PORTISHEAD	PORTISHEAD	MALTA
2 (and 2A)	CAPE TOWN	CAPE TOWN	-
3	MAURITIUS	MAURITIUS and BOMBAY	VIZAGAPATNAM
5 (and 5A)	WELLINGTON	IRIRANGI and AWARUA	-
6	VANCOUVER	VANCOUVER	-
7	SYDNEY	SYDNEY	DARWIN

8	SINGAPORE	SINGAPORE and HONG KONG	-
9	HALIFAX (CAMPERDOWN)	HALIFAX	-

Traffic for a ship in any area is normally transmitted by the appropriate area transmitting station but special arrangements may be made for traffic to be transmitted by a designated area transmitting station.

Area stations are linked by a Naval point-to-point radio network and traffic is relayed from one to another without extra charge. Thus a radiotelegram to a ship in an area other than Area 1, routed via Portishead Radio, will be forwarded to the area transmitting station to which the station is listening. Similarly, a ship may send traffic to a receiving station in any area, either directly or via another area receiving station.

Navigational warnings and weather messages are transmitted at specified times".

In addition, ships were advised to forward their latest positional information to the Ships Bureau at Portishead;

"In order that traffic may be routed to the appropriate Area Station it is essential that ships engaged on international voyages should furnish their positions and Area watchkeeping arrangements in the form of TRs:

(a) When entering or leaving port, including intermediate ports, and

(b) When changing Areas

It is important that 24 hours' notice should be given whenever a ship intends to change, or cease watch on Area Stations; failure to provide this may cause delay to traffic".

1962 saw the Portishead transmitter site expand with delivery of 6 STC 8 kW QT3 transmitters, which brought the total complement of transmitters on the site to 19.

Probably the most important development of the 1960s was the introduction of the radio telex service. Communication by this mode had been confined to contract service, but modern equipment manufactured by Phillips of the Netherlands, made a public radio telex service a viable proposition. Installation and testing of the equipment at Portishead commenced in early 1965, with the service becoming fully operational on 1 May of that year. An area at the end of 'A' wing was adapted for use as Portishead's radio telex service centre.

Initial traffic figures were very low; not too surprising considering the very few seagoing vessels fitted with suitable equipment, mostly Shell tankers. However, after a few months of operation and familiarity with the service improved, traffic levels began to rise steadily, with more and more vessels being equipped.

Each ship was allocated a 5-digit radio telex (or SELCALL - SELective CALLing) number, allocated by the ship's radio authority. This number would be exclusively issued to the ship and would only be changed if the ship changed flag or registration. Ships would make their initial call to Portishead on radiotelegraphy or radiotelephone to arrange a radio telex link on one of the numerous channels (call signs GKS, GKN and GKO) available. These channels were duplex channels, i.e. transmitting on one frequency and receiving on another. Once communication had been established, the ship could request a direct connection to any telex number world-wide. Numbers had to be dialled manually by the Portishead operator and the landline and radio circuits connected once the desired telex answerback had been received. In case of line problems or telex number engaged, it was possible for Portishead to store the telex message on a 5-unit tape for resending later; this was known as the 'store-and-forward' method, and a much improved electronic method is still used today.

Traffic to ships came in on dedicated telex lines, and a 5-unit tape was automatically produced each time a message was received. This tape was stored until the ship next communicated with Portishead and the message transmitted to the ship via a tape reader facility. Ships were made aware of messages on hand by means of an 'RTT' indicator after their call sign in the telegraphy traffic list. This slightly cumbersome but effective method of operation continued for some years before the number of ships fitted with radio telex equipment made it impractical to use radiotelegraphy/ radiotelephone to arrange contacts.

One advantage of having 'operator connect' calls through the station was the opportunity to converse directly with R/Os on board ship, many being light-hearted as Tim Strickland recalls:

View of the Portishead transmitter site, 1966.

"When working as a Radio Electronics Officer on various Esso Tankers, the primary means of contact with the ships (ignoring satcoms) was telex. Before the advent of auto-connect, you used to have to synchronise your transmissions to those of GKA and then type in something such as "GKA de GOWR pls connect me to......" The Portishead R/O would then reply with 'OK S/B' (stand by) and effect the connection. You knew you were connected when the correct answerback code appeared on your screen or printed on your paper roll. After days/months of this routine, one day, for some reason I decided on a change and when phased into Portishead telex frequency I typed "Starfleet command this is USS Enterprise sometimes known as the Esso Caledonia. Connection pls to ..." and then the number I required. Quick as a flash on my print out appeared "USS Enterprise/Esso Caledonia. GM (Good morning). This is Lt Uruha here looking particularly fine today in my tight uniform. S/B while I connect you to another galaxy far away in London town". I appreciated that for days and still dine out on the story."

The Post Office regularly produced booklets for ships' Radio Officers outlining the services provided by both Portishead Radio and the medium/short-range coast stations; the 1965 edition (which was printed before the radio telex service came into operation) makes interesting reading. For example, radiotelegrams to ships could not only be tendered by telex or telephone to Portishead Radio or any other coast station, but could also be tendered by hand at any Post Office or railway station at which telegraph business is transacted.

The long range area communications scheme was still being used in the mid-1960s, although some 'refinements' had been made since the scheme started in 1948. The 'Radio Services for Shipping' 1965 edition gives full details;

"Commonwealth countries, the Irish Republic and the Republic of South Africa participate in a scheme which provides a world-wide radiotelegraph service for ships registered in those countries.

For the purpose of the scheme, the world is divided into eight areas each of which is served by an area transmitting station and one or more receiving stations. Area stations are linked by a Naval point-to-point radio network and traffic is relayed from one to another without extra charge.

Traffic for a ship in any area is normally transmitted by the appropriate Area Transmitting Station but special arrangements may be made for traffic to be transmitted by any Area Transmitting station designated by the ship.

Each Area Station transmits traffic for ships within its own area at;

GMT 0000-0200 / 0400-0600 / 0800-1000 / 1200-1400 / 1600-1800 / 2000-2200.

However, at 0000 and 0400 GMT, Portishead transmits roll calls only; traffic for ships included in these roll calls may be obtained on request.

Traffic for those ships in areas 1A, 1B and 1C which keep a continuous watch is transmitted separately at 1015, 1400 and 1800 GMT and not included in the 1200, 1600, and 2000 schedules. Special arrangements may be made to meet the particular needs of individual ships.

Radio telegrams are transmitted once in the first transmission period after receipt at the Area Transmitting Station and repeated in each of the five succeeding periods unless acknowledged by the ship in the meantime. They are then held for acknowledgement.

Foreign ships do not normally participate in the Area Scheme and traffic is exchanged directly between ships and Portishead Radio. Traffic lists are transmitted by Portishead Radio every odd hour GMT. Most foreign ships listen to these schedules, but it would help us considerably if Owners and Agents would ask ships' Masters to ensure that their Radio Officers do so".

Some things never change. The book continues;

"When a message is received for a ship which has indicated that she has arrived at a terminal port and that radio watches have ceased, the sender is advised and no further action is taken. An exception is made for Owners' messages through Portishead Radio; they are transmitted on the assumption that the sender has more up to date information on the ship's movements. Messages for British ships which indicated that they will continue to listen to broadcasts while they are in port are treated normally and held for acknowledgement on sailing".

Ex-Radio Officer Don Mitchinson was enthusiastic about the area scheme, but with one reservation;

"I thought the area system was magic. Used to wonder how the other maritime nations managed with one. I do however think that the area system allowed radio companies to fit low power and mediocre radio gear to British ships long after our foreign colleagues had gone up-market. The Naval stations were definitely a plus with the area system. Many times off the coast of South America at night I would be offered a QSP (relay) to Portishead by Hong Kong Naval Radio. Malta Radio was also very active amongst the merchantmen".

It should be pointed out that Portishead still maintained a ships' bureau, made up of details of call signs, fittings, etc., together with the latest TR (position report) of any particular ship. These details were updated daily, and ships were encouraged to tender regular position reports and details of arrival and sailings at ports worldwide. Copies of these 'TRs' were passed to Lloyds of London via their private wire telex link to update their files.

Although many leisure craft (especially those owned by those wealthy enough to install high-frequency radio equipment) used the services provided by the British Post office during the 1950s and early 1960s, the first 'high-profile' use came with Francis Chichester's single-handed transatlantic crossing on the yacht *'Gypsy Moth III/MFDA'* during June 1962. It was a requirement of the sponsors of this voyage (The Guardian) that a daily report of around 100 words is sent, which would be relayed via the Post Office HF Radio services.

Initial discussions were held with all interested parties, and it transpired that Mr. Chichester only held a restricted radiotelephone certificate, which meant that his needs could only be provided by the use of HF Telephony.

Although the RT service was not provided at Portishead Radio at this time (calls were still be handled by Baldock and Brent RTT, although Burnham/Portishead would be the controlling centre as well as a second watchkeeping centre), it is felt details of this voyage be included for historical reasons.

Numerous items of correspondence between the GPO, The Guardian newspaper and Mr. Chichester himself took place during the early part of 1962, and looking back it was clear that there was a degree of naivety from all sides. It must be realised that this was probably the first serious and regular use of the long range service by a small craft. One interesting suggestion regarding suitable aerials was that a kite aerial is flown from the vessel – this was (thankfully) not implemented.

Backstay aerials were installed, which were satisfactorily tested.

In the event a Marconi 'Kestrel' transceiver was utilised, which would allow communication on the 4, 8 and 12 Mc/s bands. There was also a slow speed Morse option in case of telephony problems.

A chart, giving the best frequency to use at various times and positions was produced by the Post Office, and arrangements were made with the BBC to record some of the calls if they were of sufficient quality.

The yacht left Plymouth on schedule on 1st July 1962, and daily reports were successfully filed on a daily basis and published in 'The Guardian' as required. Communications were, however, sometimes difficult, and much of the transmission was relayed by Post Office operators. Despite these problems, however, daily contacts were established, and the vessel arrived in New York on July 5th, greeted by the liner *'Queen Elizabeth'* in New York Harbour.

A memo from Don Mulholland, on behalf of the Officer-in-Charge of Portishead Radio, reads;

"As you will know, Mr. Francis Chichester successfully completed his transatlantic crossing in *'Gypsy Moth III'*. There was no cause for anxiety at any time, and emergency communication arrangements were not brought into force.

Contacts between the yacht and Baldock/Brent were pre-scheduled and were, generally, satisfactory. Every schedule was covered by this station (Portishead) on the Marconi HR24 receiver but it was not necessary to offer assistance to Brent RTT.

Mr. Chichester had some difficulty in keeping his batteries charged, and at times he reported their S.G. (Specific Gravity) as low as 1160-1165. Despite this, his signals were extremely loud and compared more than favourably with signals received from transatlantic liners during the HF R/T tests".

So – a pioneering voyage in many ways, and one which generated valuable publicity to the Post Office Maritime Radio Services, and also one which made the services attractive to the leisure market.

Following on from the above successful sponsorship, 'The Guardian' contacted the PO Maritime Services again during 1963 with regard to an expedition to Greenland and Jan Mayen Island, to be undertaken by Dr. David Lewis, with a team of experts. Again, daily reports would be filed to the newspaper as before, but this time use of the 'Telephoto' service would be made. This would transmit, over radio, copies of photographs taken by a Polaroid camera which would then be reproduced using equipment held at the 'Guardian' office.

The Post Office had undertaken a few tests of this facility during 1961 from the Fishery Research Vessel *'Sir Lancelot'*, which proved successful, but this would be the first 'high-usage' of the service.

The catamaran *'Rehu Moana/MGOY'*, provided once again with Marconi Radio 'Kestrel' Equipment, set sail from the River Thames on 4th May 1963, and watchkeeping for the vessel was divided between the UK Coast Radio Stations and the Coast Stations located in the Faeroe Islands, Iceland, Greenland, and Jan Mayen Island.

Outside the range of these stations, radiotelephone calls would be made to Brent Radio Telephone Terminal as during the *'Gypsy Moth III'* voyage.

Initial radiotelephone and telephoto calls from the vessel proved successful, and the Post Office issued a press release on 7th June 1963, which read;

"At 5:30 pm on the 7th June, the catamaran *'Rehu Moana'* successfully transmitted a photograph to the Guardian office in London over the ordinary ship-to-shore radiotelephone service of the Post Office, and using the ordinary telephone system.

This is the first occasion on which a photograph has been transmitted from a ship at sea in this way. The experiment is even more remarkable in being from such a small ship as the catamaran and using only 75 watts of transmitted power – about the same as a single electric lamp bulb."

This historic transmission was made via Humber Radio, and although the quality of the picture was not great, subsequent transmissions via Stonehaven and Wick Radio were more successful.

This successful effort marked an important step forward in the Post Office policy of extending to ships at sea all the services available to telephone subscribers ashore.

Unfortunately, the catamaran lost her mast in a gale on 20th June, but a temporary aerial system was erected, ensuring daily communication was maintained. A temporary mast was rigged in Sedyisfjord on July 4th, but due to time lost and poor weather conditions, Dr. Lewis decided not to proceed further than the northern tip of Iceland, and sailed back to the UK with a new jury rig. The vessel arrived back in Stornoway on 26th July.

Meanwhile, procedures and equipment at Portishead Radio continued to evolve. A document written by the station manager of the time, T.N. Carter, describes the station's operation during the mid-1960s;

"Of the 32 HF receiving positions, six are fitted with facilities for 'search point' work, two are fitted with radioteletype (radio telex) receiving equipment, and six are fitted with equipment for undulator tape reception of high-speed Morse. Any HF aerial can be selected at any HF position and facilities are provided at each position for selecting up to 40 transmitters. Sidetone is provided via the send-receive switch at each position to avoid the possibility of two operators attempting to key the same transmitter at the same time.

A telephone intercommunication system provides communication between Search Points, Search Points and working points, and any receiving bay and the central traffic control position known as 'circulation'. Search Point positions are also provided with a special signalling system, which indicates whether their working points are free or engaged, and an additional facility which permits any signal to be extended from the Search Point receiver to the headphones of any working point operator.

'Circulation' is situated in a separate control room, and acts as an enquiry point as well as holding all Area 1 traffic and foreign ships' traffic. Other positions in the control room are Ships' Bureau, at which all shore-ship traffic is appropriately routed after reference to the extensive records of ships' movements, and the 'Finished Check' position, at which all cleared traffic is finally scrutinised before being brought to account in the separate accounts room.

From-ship traffic is carried to the control room by a system of conveyor belts and another conveyor system carries traffic in both directions between circulation and the landline positions".

You will see that the system used is very similar to that operated in 1948, but with suitable 'enhancements'; these were to continue throughout the 1960s and 1970s. The landline room, however, had changed dramatically since 1948; the article continues;

"Six landline circuits are connected to the TAS (Teleprinter Automatic Switching) system of the inland telegraph service, and a further six positions are fitted with telex equipment providing direct contact with all subscribing shipping interests. A further six circuits provide landline communication with Whitehall W/T (Wireless Telegraphy) station for the passing of traffic to and from the overseas area stations; this traffic is now largely handled via a tape automatic relay centre at Mauritius and, for this purpose, the traffic has to be appropriately processed by the originating station. In addition to the circuits already mentioned, two 'private wire' circuits give communication with Lloyds of London and the Meteorological Office to ensure minimum delay on shipping intelligence and weather messages respectively.

One side of the landline room is devoted to radio broadcast work and four positions are provided for this purpose. Each of these positions is provided with a perforator and Wheatstone transmitter (for taping and signalling by tape, respectively) as well as a Morse key, receiver, transmitter-selecting facilities and monitoring indicators".

The 'Wheatstone Machine' at the station was used to 'punch-out' messages for ships. Telegrams were prepared by using three sticks to hit one of three buttons on the machine. The button on the left made a 'dot' the one on the right a 'dash' and the centre one made a space on the tape. The skilled operator could 'type' faster by crossing over one's hands for the letter 'S' (three dots) rather than using the same hand to hit the same button three successive times.

In the mid-1960s, the then Officer-in-Charge, T.N. Carter, wrote a brief article for ships' Radio Officers, outlining the calling procedures preferred by operators at Portishead. Remember that at the time, it was not uncommon for well over 250 vessels to call each hour, and although the instructions may seem amusing now, at the time they were extremely useful;

"The quality of service we can offer depends not only on the staff and methods at Portishead Radio but also on ships' Radio Officers. In general, we find the standard of operating very high indeed, but a brief summary of some of the factors involved may help the minority of ships' Radio Officers who encounter difficulties, perhaps due to lack of experience. The experienced ship's operator will make a careful choice of his calling band by using his theoretical knowledge, by consulting the optimum frequency guide which we broadcast each Sunday, and by listening to our transmitters. If two bands appear to be usable, the higher will generally provide us with a better signal but the experienced operator will also take into account the congestion on each band as indicated by our signals".

The Landline room. The conveyor belt used to carry radiotelegrams to and from the Control Room conspicuous in the centre of the photograph.

The article continues;

"Our rate of working ships would be further increased if the many forms of superfluous signalling could be reduced. A busy search point operator cannot neglect other ships by listening to a long series of Portishead call signs nor can he offer an interrogative 'de' to the ship which pauses to listen through only sending our call sign. Initial calls should be brief and, apart from call signs, the only other information they should provide is the ship's working frequency. Calls should also be reasonably spaced because the Search Point may have logged more than one ship calling and may also have a brief wait before obtaining the use of the transmitter".

Don Mitchinson looks back with fondness at this particular era;

"I used to marvel at the reams of beautifully sent broadcasts of traffic. My imagination was stretched to the limit at the thought of five or more operators keying the 12 MHz transmitters. No matter how busy they were, at the end of transmission I would wait, and get, the eventual 'dit-dit' – thanks and so long. Happy days".

His sentiments were echoed by Tony Millatt, Radio Officer with the New Zealand Shipping Company between 1962 and 1966;

"My initial impression of Portishead was that it was very efficient, but to a young first trip Radio Officer it was a bit intimidating. Being initially on cargo ships, we did not have a lot of traffic, except for the weather reports (OBS) every 6 hours. Two of these came at the end of our watch, so we were motivated to clear them as quickly as possible and get to bed.

It was then that the fun started. You would pick out GKL on 8 Mc/s, GKG on 12 Mc/s, or whatever seemed strongest. It was never that strong and always seemed busy. There were, of course, other ships, all with the same problem, trying to get through. We only had 100 watt Marconi Oceanspan transmitters, and the others would often get in first. It was bad practice not to clear the OBS though (i.e. the second mate would have words to say), but they would invariably go in the end.

In those days we had the area system, and once clear of Panama on the way out, we would have a quick flirtation with CKN Vancouver (never easy) and then transfer our allegiance more permanently to ZLW Wellington and ZLB Awarua. Occasionally in the Pacific, we would hear Portishead and clear traffic directly, but not often".

Let us not forget the radiotelephone services; although not yet operated from Portishead Radio directly, it is interesting to look back on what the service was like during the 1960s. An article written by R.M. Billington and published in the September 1963 issue of the I.E.E. magazine, describes the service in full;

"The present service for ships provided by Rugby (transmitting) and Baldock (receiving) caters for handling calls to ships two at a time. Inverter privacy is available if required, and the service is open 24 hours a day although, of course, adverse propagation conditions sometimes limit the period during which a satisfactory circuit can be set up with a distant ship.

The number of calls has not changed very much over the years, although since 1960 the MF service has handled several thousand calls from long-voyage passenger ships in home waters, a number of which might otherwise have been taken on the HF service.

A telephone call to a passenger on a ship can be booked by asking the local telephone exchange operator for 'Ships Telephone Service'. When connected to the controlling operator in the International Exchange, the caller asks for a ships' radiotelephone call and gives the name of the ship and the person on board required. This information is passed to the Radio Telephony Terminal, which sets up the radio circuit. Calls from persons on board ship are initiated via ship's Radio Officer. Contact between the ship and the Radio Telephony Terminal may be established in several ways; a schedule of meetings may be pre-arranged; or a meeting may be arranged by telegraph via Burnham; or a call from a ship may be received by means of a listening watch at Baldock or by means of an automatic calling system actuated, by auxiliary radio receivers, by the ship's radio carrier.

A family of directional aerials is available at both Rugby and Baldock to suit the position of the ship, and there is a choice of five HF bands to suit propagation conditions. Telephony, of course, requires a higher grade of circuit than telegraphy, and care is taken in the selection of aerials and the choice of frequencies.

The transmitters at present in use at Rugby have a peak envelope power of 30kW and are suitable for single and double sideband emission over the range 4-27.5 Mc/s. Transmitter tuning is preset, and each transmitter can be brought up on any one of six frequencies, selected by remote control from the station control centre, in about a minute. A transmitter/aerial matrix gives access from either transmitter to any one of ten double-ended rhombic aerials, spaced at 18-degree intervals.

At Baldock, the receiving station, two independent-sideband receivers are used to receive the signals from ships. If the ship's emission is double sideband, the sideband least affected by interference is selected. A wide selection of directional receiving aerials is available.

Rugby and Baldock are connected by landlines to the Post Office Radio Telephony Terminal in London. The Terminal contains the engineering control positions and terminal equipment such as voice-operated singing suppressors and inverter privacy equipment. The radio circuit between ship and shore is extended to the International Telephone Exchange for connection to the inland telephone network or extension to other countries overseas. The telephone control of the circuit, charging of the calls etc. is done by the International Exchange".

Returning now to Burnham, and specifically the to-ship telegram service, all to-ship traffic from the Admiralty network and telegraph offices came in on TAS machines with the text appearing on paper tape (approximately 5/16" wide). This tape had to be cut to length and gummed to telegram forms before being passed to the ships' bureau and circulation. Not surprisingly, there have been plenty of stories to recall about incidents to do with these incoming messages; one which has always been locally popular is the Radio Officer (who shall remain nameless) who proceeded to gum the wrong side of the tape of a very long telegram, resulting in bits of tape falling off the telegram form on its way to the Ships' Bureau. Another incident occurred when a new recruit was told to dip the whole telegram in a container of glue to 'save time' - this resulted in a large sticky mess being transported down the conveyor belt accompanied by the collective mirth of the landline wing.....

Keyboard skills were naturally required to use the telex and TAS machines; indeed, when sending telexes on the latter, there was no local copy, so one had to be sure that what one keyed was correct; luckily, the telex machines had a paper copy and a 5-unit tape reperferator facility, so long telegrams could be checked and corrected before being sent to line.

Portishead Radio Neville Edwards recalls:

"One of the things we had to learn in the training school was touch-typing. This was an absolute must when using a TAS machine. These machines were similar to a telex machine but there was no printed page - you could not see what you had typed - hence accuracy at touch-typing was paramount. The incoming side was on a tape with a sticky back - this was then moistened and affixed to incoming telegraph forms. At one time this was how all telegraph offices would have worked. An art form at the time to moisten the tape - no too much water - using a special metal container with a brass roller. Take tape in both hands and move over wheel. Affix to page and trim off with a special cutter on your finger, ensuring of course you did not cut a word in half. There used to be mountains of tape from all over the UK and incoming from overseas offices via International telegrams in London."

The traffic list was generated by the use of a telex machine in local mode, with the traffic list officer typing the call signs (in alphabetical order) of ships for whom traffic was held. A 5-unit tape copy was produced which was fed through a tape-to-Morse code convertor for broadcasting. Traffic lists for British-registered ships were broadcast each even hour, and lists for foreign-flag vessels each odd hour.

At the end of 1965, Shell International Marine carried out a survey with regard to reception of Portishead Radio in the Far East, comparing signals with those of Scheveningen Radio (PCH), Rogaland Radio (LGB), and Norddeich Radio (DAN). The results make for intriguing reading, especially the differing conclusions offered by Shell and the GPO.

On 1st December 1964, Shell arranged for all charterers' business to be passed through Portishead in order to ensure consistency in operational procedures for all vessels, both UK and foreign flag. Considerable difficulty was encountered with foreign-flag vessels, which it was felt was due to the different operational methods used at Portishead when compared with those used elsewhere, and of course to the heavier traffic load borne by Portishead.

The GPO produced a booklet explaining its procedures which was sent to all Shell vessels, and also set up an additional Search Point. These actions resulted in a substantial improvement in performance, and the majority of messages were cleared adequately through Portishead from about May 1965 onwards.

Many ships, however, still claimed they could not raise Portishead at times when their own National coast station was readily workable. As a result, many messages to these ships were duplicated, one being sent via Portishead, and the other via the appropriate National station. At first, it was thought that this was again due to operational misunderstandings, but careful enquiry suggested that there was at least an element of truth in the idea that Portishead was in some way inferior. It was therefore decided to conduct a simple test to try and prove the point, undertaken on the vessel *'Serenia'*.

W/T Operating wing, mid-1960s.

The results of these tests suggested that when propagation conditions were good, from 0900 to 1300 hours, there was not an enormous difference in the availability of the four stations. During the marginal times, Norddeich showed a big advantage over all others. Scheveningen was nearly as good, while Rogaland and Portishead were on a par. The day-by-day figures show that sometimes Portishead provided the better signal, but that these occasions were far outweighed by those when the reverse was the case.

Shell concluded that: "There can now be no doubt that the service provided by Portishead is substantially inferior to that from Norddeich or

Scheveningen. It is not so markedly different from that given by Rogaland. It would appear then, that Portishead's deficiencies are due to a combination of heavy loading and poor 'radiating efficiency'."

Landline Room, mid-1960s. Broadcast area top right.

Whilst these tests may have had some merit, it was felt in Post Office circles that the Shell Radio Officers (many of whom were Dutch or German) tended to prefer their own station anyway, and also, serving on foreign-flag vessels, could not avail themselves of the 'area scheme'. Another aspect that Shell raised, was the non-use of 'callbands' which automatically sent call signs when a transmitter was not in use. These were transmitted when the circuits were 'idle' to enable the ship's Radio Officer to tune in to the strongest or most readable signal. Again, it was felt that unfamiliarity with the Portishead Radio operational system may have contributed to the apparent problems. Nonetheless, an interesting exercise for both parties.

It may be interesting to review the terms and conditions for staff employed at the station in 1965.

The basic scales of pay were:

At Age 21	£651
At Age 22	£684
At Age 23	£717
At Age 24	£750

At Age 25	£789
Then by annual increments:	£827
	£860
	£899
	£943

Annual leave was initially 3 weeks 3 days, thence:

After 10 years' service	4 weeks 2 days
After 20 years' service	5 weeks 2 days
After 30 years' service	6 weeks 2 days

The working week was defined as a 43-hour working week, excluding meal breaks. The general pattern was described as below, but could be different for each individual duty.

Day	Duty	Hours	Total
SUNDAY	Rest Day		
MONDAY	Evening Duty	5 pm – 11 pm	6 hours
TUESDAY	Afternoon/Evening	1 pm – 11 pm	9 hours 30 mins
WEDNESDAY	Day Duty	9 am – 5 pm	7 hours 20 mins
THURSDAY	Morning Duty	8 am – 1.30 pm	5 hours 30 mins
FRIDAY	Morning Duty	8 am – 1.40 pm	5 hours 40 mins
	Night Duty	11 pm – Midnight	1 hour
SATURDAY	Night Duty	Midnight – 8 am	8 hours

TOTAL HOURS   43 hours

On February 22nd, 1966, a new service for ships was introduced whereby crew members could send SLTs or telegrams to Vernons Pools giving their selected football coupon numbers in blocks of weeks. This was of course in the days before the National Lottery when a pools win made you rich beyond your wildest dreams.

Apart from the Vernons Pools messages, SLTs were a quick and cheap way of ordering flowers via Interflora and ordering gifts via Kays Catalogue. Arrangements were made with these companies to send all SLTs to them in bulk on a daily basis to reduce costs. To give an idea of how popular these SLTs were, the number of Interflora SLTs handled by the station in the early 1960s were:

1963 – 46,814

1964 – 43,296

1965 – 41,320

This service became extremely popular, although the reception of numerous coupon numbers could be tedious at times. The service ceased during the 1990s when the popularity of the football pools began to wane with the introduction of lotteries and similar schemes.

Developments in the short and medium range service continued unabated; on 1st April 1966, a new type of radio station was opened at Fairseat, Kent. This comprised a VHF sending and receiving station remotely controlled from a main coast station, in this case North Foreland Radio. This was the first of what was to become a whole series of remotely controlled VHF stations around the United Kingdom coast.

Mention should be made at this time of Sir Francis Chichester's epic single-handed 'Round the World' voyage on *'Gypsy Moth IV'*.

The first item of correspondence came from Mr. Chichester on 28th April 1966, in which he gives details of his projected voyage, details of the equipment (again, a Marconi 'Kestrel' transceiver), and details of the vessel (*'Gypsy Moth IV'*, call sign GAKK).

The actual voyage commenced from Plymouth on 27th August 1966, with the first schedule with the 'Sunday Times' at 1700 GMT through Brent via Land's End Radio. It was intended to use Brent RTT for as long as possible, but schedules were made with other National Coast Stations such as Cape Town, Durban, Perth, Sydney, Wellington, Buenos Aires, and others). This would of course mean that Post Office involvement would not be as great as the earlier transatlantic crossing of *'Gypsy Moth III'*, although of course progress would be monitored with interest.

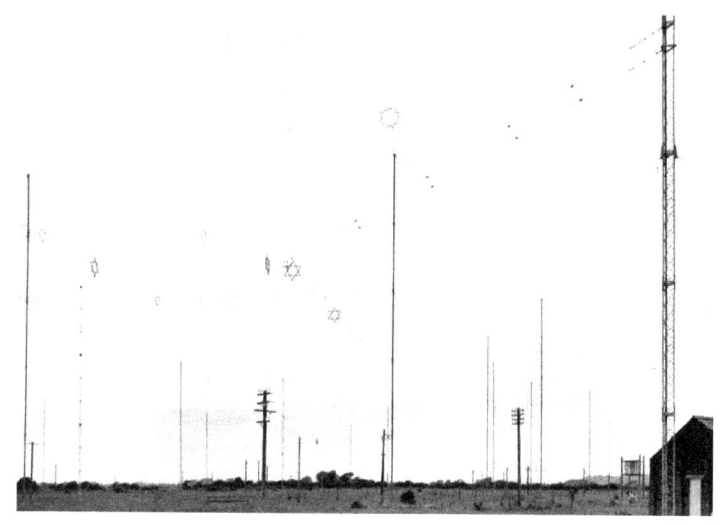

The receiving aerial farm at Highbridge, 1965.

Mr. Chichester made it clear that he would not accept calls from any source apart from Lady Chichester. A note from him states that 'if any person other than Lady Chichester attempts to book a radiotelephone call to *'Gypsy Moth IV'*, the booking should be refused giving the reason as the radio equipment is supplied from batteries and use must be limited to obtaining information for navigation purposes or for any emergencies that arise'.

There was immense public interest in this voyage, and when the vessel returned to Plymouth on 28th May 1967, Land's End Radio reported an all-time record month for link calls. The Officer-in-Charge at the time notes 'Press and television companies had ten chartered ships passing traffic through the station. The 'Daily Mail', 'Daily Express' and 'Daily Telegraph' send over thirty pictures by radio on 2 M/cs to their offices. Most of them were successful, and some of them were published. BBC and Rediffusion TV did much of their recording on 2 Mc/s and Sir Francis had six very good calls through the station'.

Numerous bouquets were received from the media; The BBC Outside Broadcast Unit, The Daily Express, The Daily Mail/Daily Sketch, and Rediffusion Television all sent glowing commendations to the Post Office Maritime Service. In all, 231 calls were made via Land's End Radio, with a few other calls handled by Niton and Ilfracombe.

The use of the Long Range service during the voyage is unfortunately not documented, although there is no doubt that use of the service was made regularly whenever possible.

As a result of the great media attention following this epic voyage, a single-handed round the world yacht race was organised for 1968/1969. This race became famous not because of who won it, but because of who didn't. The name of Donald Crowhurst was not known before the start of the race, but it became synonymous with this particular race.

Anyway, from the beginning, it became clear that the Post Office maritime radio services were going to be heavily involved following the success of previous yacht communications. An internal memo dated 1st April 1968 states:

"A single-handed yacht race around the world will take place during the latter part of 1968. Contestants will commence from Plymouth any time between 1st July and 31st October 1968 and must pass around Cape Good Hope and Cape Horn, finishing at Plymouth sometime in 1969. The 'Sunday Times' newspaper is sponsoring.

As is usual in these events, there will no doubt be a demand for R/T Communications. The greater majority of this will be on the H/F R/T services and on the outward passage".

This race attracted numerous entrants, and this history of this race is well documented elsewhere. However, it is interesting to look back on some of the correspondence dealing with some of the entrants. For example, Commander King who entered the race on board the *'Galway Blazer'* was due to attempt the voyage non-stop, sponsored by the Daily Express, and fitted with the now customary Marconi Kestrel II radio equipment. Robin Knox-Johnston, on the 'Suhaili' (sponsored by the Sunday Mirror) also was equipped with the modified Marconi Kestrel II unit allowing 16 Mc/s communications. In fact, the first provisional list of entrants showed that all but two were equipped with this equipment, the others having SP Radio 'Sailor' equipment.

Again, the Post Office provided a list of frequencies not only for the UK Radiotelephone Service, but also for other National Coast Stations worldwide.

However – a letter from Donald Crowhurst, who entered the race on board the *'Teignmouth Electron'*, indicated a somewhat different attitude to radio communication;

"I am a competitor in this event and would like, if at all possible, to use telegraphy during the event not only because of the improved intelligibility possible, but also because of the tremendous improvements in efficiency that derive from the use of transistor class C stages. With electrical energy at a considerable premium I feel sure that I do not have to labour the arguments for my reasoning".

Basically, what he requested was authority to conduct all radiocommunication by Morse code with Portishead, rather than making radiotelephone calls via Brent RTT. The reply from the Post Office was sympathetic but called for the appropriate operating licence to be obtained, and for the radio equipment on board to be type-approved.

In the event, such approval was indeed forthcoming, and the *'Teignmouth Electron/MZUW'* set sail on 31st October 1968, on what was to be an eventful voyage for all concerned.

Full details of the voyage can be read in the excellent book 'The Strange Voyage of Donald Crowhurst' but suffice it to relate here that his position reports received at Portishead (in morse code – his R/T equipment for communication with Baldock Radio did not function properly) did not always ring true. It was during his regular position reports that suspicions were raised; the positions given by Crowhurst put him in the Indian Ocean, even though his signals were being received on aerials directed to the South and Southwest. The race organisers were also unsure about some of the positions being reported, as in some cases the yacht would have travelled over 200 miles each day, not a realistic option taking into account the size of the vessel and prevailing weather conditions. In reality, Crowhurst was circling in the Mid/South Atlantic as he became aware that the yacht would not survive the conditions in the Southern Ocean. Sadly he was to take his own life, abandoning the *'Teignmouth Electron'* but leaving his log books on board. His abandoned yacht was found on 10 July 1969. Films such as 'Deep Water' and 'The Mercy' have attempted to keep the episode in the public eye, and television documentaries on the subject are regularly screened.

Meanwhile, the transmitter site at Portishead took delivery of a third transmitter, which was reported in various engineering journals:

"The third short-wave transmitter which has recently been constructed at Portishead Radio by the Post Office Radio Branch, has been provided to replace the original short-wave transmitter which has now become rather out of date in many features. The original transmitter was first installed in 1927 at Devizes, and when it was built the frequency tolerances allowable were not as rigid as they are at the present day. The transmitter gave excellent service with the comparatively few ships which were, at that time, equipped with short-wave apparatus.

As the merits of short-wave radio communication were gradually appreciated by the ships passing their traffic through Portishead, so the volume of traffic on those waves grew, until, in 1931, the first transmitter became unable to handle the increased traffic. A second transmitter, incorporating all of the latest improvements, such as crystal control of frequency was therefore installed and put on traffic in 1932.

As more and more short-wave services were being operated all over the world, the frequency tolerances became rather more stringent than could be met conveniently by the original transmitter. It was therefore decided that, rather than apply crystal control to the old transmitter, a new and completely up to date transmitter should be constructed, the old set being retained as an emergency transmitter. This scheme would give Portishead two modern crystal-controlled transmitters, with an emergency transmitter that could be readily tuned to any of the frequencies used by the two main transmitters."

1968 saw the introduction of the 'Daily Telegraph' press service to the '*QE2*', using a Plessey PT200 system and/or a 'Piccolo' system. The former system used a 6-unit code which would key a Portishead transmitter, with the ship's equipment automatically seeking requests/repeats via a telex 'ARQ' path through Somerton radio. The 'Piccolo' system used a separate Portishead transmitter which was modulated in USB (upper side band) by 32 audio tones in the range 320-650 cycles/second, each tone representing a 5-unit symbol.

An announcement was made to the staff on 18th September 1968 with the following notice;

"For your information, additional equipment is being installed in the Radioteleprinter Room for engineering tests on the practicability of setting up a special press service between the Daily Telegraph and the *'QE2'*.

The basic proposition is that the use of 'varitype' machines with the standard GPO datel modem units, and the use of a six-unit code, may allow transmission of a tape which, on board the ship, could be translated into direct printing of a newspaper in standard newspaper format.

It should be emphasised that, at present, the proposed system is only being subjected to engineering tests. If these experiments are successful, Cunard, 'Daily Telegraph' and the GPO would next need to consider costs and charges and only then would we be seriously involved in operational procedures, staffing, etc."

This service commenced trials on 24th December 1968, with the equipment being located adjacent to the radio telex area at the end of 'A' wing. A staff notice issued on the 19th December 1968 gave details of the operation;

1. At, or shortly after, 0145 GMT each day, radio teleprinter contact should be established with the ship, using normal radio telex service methods. The ship and Portishead should change frequency as necessary to ensure that the best possible communication conditions have been established. The available frequencies are;

PORTISHEAD RADIO	SHIP
4316.85 kHz	4169.0 or 4171.5 kHz
6397.85 kHz	6248.5 or 6254.0 kHz
8582.45 kHz	8337.0 or 8341.0 kHz
12714.85 kHz	12487.0 or 12502.0 kHz
17137.65 kHz	16645.0 or 16653.0 kHz
22527.85 kHz	22171.0 or 22180.0 kHz

During this initial exchange of signals, Somerton should be asked to confirm that they are fully satisfied with their receiver tuning and that the AFC (automatic frequency control) facility is not in use.

2. Immediately before 0200 GMT the ship should be asked to stand by for Plessey and Piccolo signals and the Burnham operator should;

(a) Request Portishead to bring up the 'Piccolo' transmitter on whichever of the following frequencies is in the band established as optimum:-

i. 4314 kHz   6395 kHz
ii. 8580 kHz   12712 kHz
iii. 17135 kHz   22525 kHz

All in upper side band (USB) mode.

(b) Ensure that the Modem 2A unit is switched on (as indicated by the lamp on the face of the unit when power is provided from the power point on the wall);

(c) Place the switch on the small unit (alongside the modem unit) to 'STD';

(d) Use the telephone associated with the modem unit (by lifting the handset, pressing the button marked 'tele' and dialling the Daily Telegraph office to advise that 'the circuit is ready for their transmission using the 4-wire method';

(e) Replace the telephone handset on the table (not back on its cradle);

(f) Place the switch on the small unit (alongside the modem unit) to '4-wire';

(g) Depress the button, on the telephone, labelled 'send data';

(h) Place the Autospec/Datel switch, on unit A at RTT position A to 'datel';

3. During the transmission the signals should be monitored, at intervals, using the facilities at RTT position A. Using the local receiver it should be possible to monitor;

(a) On the Portishead frequency listed under 2(a) above, the musical sounds of the Piccolo transmission;

(b) On the Portishead frequency listed under 1 above, the FSK signals of the Plessey transmission, but it should be borne in mind that these may cease at intervals when the ARQ path is activated.

(c) On the ship frequency listed under 1 above, the ARQ path; a high audio tone will be heard when the ARQ path is inactive but, at intervals, this will change to a lower audio tone whilst repetition requests (RQs) are being sought".

The Somerton lines were also monitored to ensure that no problems were encountered; back-up measures were also employed in case of landline and radio path problems or failure.

This system was used for several years by the Daily Telegraph until superior technology made this somewhat involved method redundant.

In addition to the above, a direct printing telegraphy (radio teleprinter) service was brought into use at the end of the 1960s. A system known as SITOR (Simplex Teleprinter Over Radio), produced by Philips of the Netherlands was provided at selected coast stations, permitting telex-type communication between ship and shore. It was also used for exchanging radiotelegrams between ship and coast stations, with the coast station also having the facility to store and relay the message when direct connection between the ship and the shore subscriber is not practicable.

Initially, it appeared that those in authority thought the new service would not catch on. Marconi engineer Ron Stringer recalls;

"A major part of the work of my group involved the pre-sea testing and into-service introduction of new products. When the ITU eventually agreed an international standard for maritime Telex over Radio (TOR) I was responsible for proving and fitting the early Marconi 'Spector' terminals. As these were taken up by more and more shipowners, it was found that British shipowners felt inhibited by the lack of facilities offered by the UK long range coast station. Yes, the Swedes and the Dutch PTT were offering automated and semi-automated HF TOR facilities, but the London-based owners were used to sending all their to-ship traffic via Burnham. Could those foreigners be trusted with the traffic from British owners? Wasn't it more expensive to use foreign stations?

Marconi Marine pointed out that the Dutch had 16 HF TOR terminals in use at PCH and they pressured the GPO, directly and via the Post Office Users' Committee, to install TOR equipment at Burnham. Not necessary, was the reply. Their traffic statistics showed that in the UK, HF Morse W/T traffic was on the increase and this new-fangled telex stuff was not growing at all. A new building programme for Highbridge was to be implemented and I remember Jack Wall proudly announcing to the POUC that, although there were no plans to provide TOR, there were to be 100 new W/T operating points provided. The fact that the TOR couldn't grow at Burnham because there were only very limited facilities offered, was ignored. As was the fact that W/T traffic was discouraged at many HF coast stations in favour of TOR. The increase at Burnham came from the many small, low-traffic FOC ships that were being shunned overseas; the high traffic ships and the more progressive companies had moved to TOR. Eventually the GPO agreed that they would accept a number of Marconi 'Spector' terminals on free loan so that an enhanced service could be offered. As a result, I paid a number of visits to both Highbridge and Somerton over a period of several months. TOR traffic then began to grow via Portishead Radio. How surprising!"

A pilot service was provided in December 1968, using Marconi 'Autospec' equipment to meet the special requirements of a few ships. This system, known as 'SPECTOR' (Single Path Error Correcting Teleprinter Over Radio) was used in parallel with 'SITOR' for a time before SITOR became the international standard. Portishead had equipment of both types until the introduction of the automatic radio telex system in July 1984.

Encoders for selective calling, in accordance with the internationally agreed system, were installed at all coast stations. This greatly facilitated making contact with ships which hitherto had been dependent on human watchkeeping. Impetus was given to the adoption of direct printing and selective calling by the shipowner's desire to be able to use the ship's Radio Officer to maintain other electronic and electrical equipment on board ship, and by the review of distress watchkeeping by ships being made by IMCO.

At the start of the 1960s, 143 operating staff and 4 maintenance staff were employed at Burnham, with 14 transmitter engineers at the Portishead site. At the end of the decade, staff numbers had increased accordingly with the amount of traffic handled. However, it was clear that staff numbers remained insufficient taking into account the increase in business, and an exchange in Parliament between John Stonehouse (Postmaster-General) and Ray Dobson (Union of Post Office Workers) from April 1969 is recorded in 'Hansard';

Mr. Dobson asked the Postmaster-General whether, in view of the staff shortages at Burnham Radio, he will give an undertaking not to withdraw the ship letter telegram service from ships.

Mr. Stonehouse: At present it is not my intention to suspend the service.

Mr. Dobson: While I am grateful for that intimation, may I ask my right hon. Friend to go further? He must be aware that there is a drop of commitments in this station due to staff shortages. How long will he allow this to go on before he has to think seriously about withdrawing the services for ships at sea, which are very important indeed?

Mr. Stonehouse: I agree about the importance of this situation and I will keep it under close review.

Mr. Dobson asked the Postmaster-General when he proposes to start new negotiations over the regrading of Post Office radio operators; what is the present staff shortage; and what his estimate is of the staff shortages in three and six months' time, respectively.

Mr. Stonehouse: Negotiations have been going on for some time. There are at present some 30 vacancies. I cannot estimate what the position will be in three or six months' time as the recruiting position fluctuates from month to month.

Mr. Dobson: While my right Hon. Friend is aware that the recruiting situation fluctuates from month to month, is he aware that it takes about six months to get someone in to fill vacancies? As the shortage of staff is in part due to lack of progress over regrading proposals which are being held up by his Department, will he press for the settlement of these as soon as possible?

Mr. Stonehouse: I do not accept that regrading proposals are being held up in our Department. There has to be some very close consultation with the staff sides.

It is an interesting fact that Ray Dobson quoted above was an ex-Radio Officer at Portishead Radio, as was his son Roger.

Mike Howley recalls the end of the RO1/RO2 division:

"I had made approaches to the GPO to join in previous years but refused to join as an RO2, since I had a 1st Class ticket. Martin Davies was the first RO to join under the new scheme, (in early December 1969 I think). I was the second RO to join directly as an RO1, on 4th January 1970 while Martin was still in training in the school hut.

Although the duties at Burnham were combined and no RO1/RO2 division of labour remained there, the old system persisted to some extent at MF coast stations. Duties at coast stations HAD to be arranged so that an RO1 was on duty at all times. When casualty work started, the RO1 took over the W/T console, while the RO2 took over the RT console (irrespective of where they had been seated before the distress call).

In November 1970, I was posted to Anglesey/GLV as RO1, but with the proviso that I pass my German language exam within a fortnight, otherwise I would be sent back to Burnham. (They needed an RO1 there, but did not need an RO2 there.) I crammed an awful lot of study into that fortnight and somehow managed to pass the exam, and stayed at GLV until 1976.

Whilst the duties of RO1s and RO2s at Burnham were combined, the old grades persisted. ROs with 2nd Class tickets had to get their 1st Class ticket to become RO1s. When the General ticket was instituted, the PO sent RO2s on courses to Southampton to train up and get their General ticket and therefore to become RO1s. (That was in 1975 to 1977.)

In November 1970, I was posted to GLV as RO1, but with the proviso that I pass my German language exam within a fortnight, otherwise I would be sent back to Burnham. I crammed an awful lot of study into that fortnight and somehow managed to pass the exam, and stayed at GLV until 1976.

At Burnham, the main difference between RO1s and RO2s was that the latter had to work nights, whereas the former did not. After 1969, both grades had to work night duties.

RN radio operators at Burnham only worked on HF. They did none of the other tasks that GPO ROs did. Most of the time they sat on 8 MHz because that is where warships were most likely to call with ZBOs."

As the decade drew to a close, it is fascinating to read the recollections of former Portishead R/O Neville Edwards, who describes the station at Christmas time;

"Before computers and other such advanced technology in the 1970s, just a mass of paper and posties doing all the carrying to and from the operators. Messages delivered from the central position and Greens/SLTs from the incoming belt. These then on to the landline. Greens, by the way, were the incoming messages. Everything with the exception of SLTs were typed on to them. Things like MSG, PDH, etc. meant nothing to us. These were just for shipboard charging. All were handled in a similar way. SLTs would be copied and the original (subject to them being clean copies) forwarded to the recipient.

The Christmas period was just manic with messages in their hundreds; no, more like thousands to be despatched to the area network for our vessels scattered around the globe and likewise incoming traffic from all points of the globe to be sorted and despatched onto the landline/telex networks. We used to have RO's on detached duty from around the coast and also a number of telex/TAS operators from inland telegraph offices. Not too many decorations around but loads and loads of cards from shipping companies, private individuals and ships strung up around the station.

The W/T side was a constant stream of ship having huge QRYs of 68 plus on 8 MHz. Not uncommon sometimes even more. Sending the then broadcasts on Area 1A/B were immense and I remember being on a supplementary broadcast which really never stopped as one pile of messages was removed and the next lot placed in front of me. Thank goodness for an auto key. Most of the monitors in the early days were old CR300s – I think or similar. There was one main console which was somewhat more advanced than that – which was the home of the main broadcast point. This was under normal circumstances where the traffic lists and all other broadcasts went from. This was situated in the old landline room which was just a clatter of noise from people talking and printers rattling away. Directly behind this position was the main Admiralty networks which consisted of, if my memory serves me correctly of two outgoing arms and about 4 incoming. This is where all the traffic to and from the area scheme originated.

The main traffic list could be punched up onto tape using a Post Office machine which was an art form in itself. This was a two-unit tape. The machine required certain combinations of letters to produce punctuation and the like. We had some who were just plain genius when it came to this machine. I could use it but it was a real chore. I would rather hand key the lists than try and type them up with this antiquated monster. The foreign list then was in a large box which was in alphabetical order of course but when the time came to send the list someone bowled over and sat down at the desk on which was a hand key and then off they went with the list – taking a box at a time and just turning the messages over to see the next call sign. OK until you had something like one of the passenger boats then a scrabble to turn them over without losing the rhythm of the list. All great fun and an art form.

In the main Control Room area was also the bureau which held all the ships in name files. With up to date or hopefully up to date info on which area station they were listening to – for UK, Commonwealth vessels. Also any info which ships had given us. The men on this point would check all incoming messages and append call signs and relevant info and then direct them to the right destination whether for us at GKA or onto the area network. So you can imagine that the guys listed to the Admiralty network had a pretty hard slog as well. It was all non-stop and at times you would think is this ever going to quieten down. I would not have missed it for anything. Like all my colleagues we took great pride in our station from then until the final days.

Back then attendance over Christmas and Boxing Day was mandatory unless of course you had a rest day. Then it was probably in to do some overtime. It was many years before it became the norm to have it off. Coverage then was by volunteers. Always plenty of them as the pay was good and days off in lieu."

CHAPTER FIVE

THE 1970S

At the start of the decade, 143 operating staff and 4 maintenance staff were employed at Burnham, with 14 transmitter engineers at the Portishead site. However, changes were afoot as the area scheme was shortly to close, and consequently the Naval presence at the Burnham station was to be discontinued. An overview of the HF long range radiotelegraph service written in 1970 explains;

"This area scheme is now closing down because most of the overseas area stations are being withdrawn from the scheme. The remaining stations at present are Mauritius, Singapore, and Cape Town. These are likely to withdraw in 1971. This means that all traffic between the UK and British ships will have to be handled by Portishead/Burnham. The estimated increase is 40% of the whole traffic (British and foreign) now handled at Burnham/Portishead. Plans have been made, therefore, to augment equipment and staff accordingly, but unfortunately, no additional revenue will accrue to the Post Office as we already receive it under the area scheme. As an interim measure, 12 additional operating positions are being provided at Burnham and 14 additional transmitters will be used at Dorchester Radio Station to handle the additional traffic. These transmitters, although not ideally suited to the maritime service, are redundant from the fixed service.

The interim scheme will tide us over the next three or four years. The 'final' scheme envisages the building of a new receiving station at Somerton Radio Station and the closure of Burnham. Additional and more flexible transmitters will be required at Dorchester, and ultimately it may be possible to give up Portishead"

This was to be the first of many attempts over the next 25 years or so to close the Burnham site down; thankfully none of them came to fruition, although it will be noted later that various other methods (apart from high traffic levels) would come into play.

In 1972 the Union of Communication Workers (UCW) representatives from the station met with county council planning officials and Langport Urban District Council to discuss housing and associated problems in the Somerton area. It became clear that Somerton would continue to be much sought after by commuters to the Yeovil and Bridgwater areas and that property would be keenly competitive and high in price compared with similar housing in the Highbridge and Burnham-on-Sea areas. Somerton also did not offer the same social facilities offered by the current location. The Union (under the auspices of Gerry Knott and Jim Byrne) therefore recommended that they would not make any recommendation for such a move until favourable terms of transfer had been negotiated. It was also clear that many staff would not be prepared to move to the Somerton area for family reasons.

The HF Radiotelephone service was transferred from Baldock to Burnham on 26th April 1970 in order to integrate it with other maritime services. Under the new system, vessels would be connected directly to the United Kingdom telephone network rather than through the International Telephone Exchange. In the to-ship direction, subscribers would simply ask the operator for the 'Ship's Telephone Service' to be connected to Portishead Radio. The ship would then be called in the traffic lists until making contact and the call taking place. Ex point-to-point receivers and consoles were utilised, similar to those used at Somerton.

However, it soon became clear that there would not be the capacity at the Burnham site to handle the amount of R/T traffic, so existing consoles at the Somerton receiving site were used, with staff being transported to and from the site daily.

Somerton R/T console.

Rugby continued to provide the transmitters, and further transmitter sites were added later in the decade. Some services were transferred to the Dorchester transmitter site in 1972 as the service expanded. The number of calls handled increased slowly, and the provision of a maritime version of 'Lincompex' to improve the quality of telephone calls was successfully made. About half a dozen British ships were equipped with this function at this time.

The new HF R/T positions at Highbridge.

Radio Officer Mike Sheehy recalls working Portishead Radio during the 1970s;

"A good deal of my time had been spent on tankers and for me the memories of Portishead Radio are strongest for the seventies - when slow steaming around the Cape up to Europoort or a lightering position off the Bahamas or in the Gulf of Mexico when a link with home was of the utmost importance. Did you really give us QRY29 (turn 29) on GKT51 or is that just a fond memory? And did we become wary when getting QRY3 only to find QRY1 was the *'QE2'* and QRY2 the *'Ark Royal'*? No, I know it's really not true. You did your utmost to open other frequencies and transfer us around until we could make a commercial link and that is what made Portishead special".

He goes on to relate;

"I was on a new bulk carrier, the *'Naworth'* (again, in the seventies) and had been on her from her trials from Cammel Lairds. I was promised a four-month trip but when the time came for relief there wasn't one, so I stayed. I ended up getting literally Shanghai'd as the ship loaded cotton for that port. We ended up spending five and a half months at anchor or on buoys in Shanghai roads before moving to another Chinese port and loading coke for

Romania (via the Cape). By the time we had entered the Med I had spent over a year on the *'Naworth'*, a lot of that time at anchor, and decided enough was enough. I called Marconi's at East Ham and asked to be relieved; they hummed and harred. I began to get a bit desperate as our next orders were for Churchill, Hudson Bay, loading grain for Poland and I could see no prospect of getting off at either of those two places. The strain must have been evident in my voice because as I was left staring at the dial of my receiver I heard the operator at Portishead say in a very calm voice 'And the best of British luck mate'. That voice did more to reassure me than anything else. I am happy to say that I did get off afterwards at Gibraltar outbound for Churchill when we called in for bunkers, twelve months and three weeks after leaving Liverpool Bay. Not the longest trip in the world I know, but one that was made bearable by that voice on the ether from Portishead Radio".

Radio Officer Claude Pereira writes from the M/T *'Verdi/9HJE'* about his experiences during the early 1970s;

"I was on the *'British Venture'* and was sending a 'Bingo' (Position) message on HF W/T to the BP Head Office London. The weather was bad, the ship was rolling and pitching badly, my sending must have reflected the state of weather! The operator asked me 'what is happening mate?' and I told him. Then he replied 'OK Mate, take it easy, take your time – I am with you'. That was real nice of him.

Another time I remember is when my English Captain wanted to put in a call (via GKT) to his wife in Newcastle, and the phone was continuously engaged. This nice guy of an operator said 'hold on mate, I will get the telephone exchange to disconnect the call and give you priority'. You should have seen the BROAD smile (we in India call it a Jeelabi smile) on the Old Man's face! That is what I call real assistance/courtesy to mariners".

Mention must be made here of the arrival of 'Sailor', the ginger-and-white radio station cat in the early 1970s. Originally employed by the Post Office (and indeed on the payroll) as a mouser, he was a regular visitor to 'C' wing, often to be found asleep on top of the receiver consoles. Unfortunately, some Radio Officers teased him unmercilessly, and any unsuspecting R/O coming on watch who tried to stroke him would more often than not get a vicious right hook complete with sharp claws to the arm! Visitors to the station would often spend more time with the cat than with the staff, much to the annoyance of the overseers and local management. He was very much a celebrity, and featured in numerous press articles about the station throughout his life.

1970 saw one of the very first trials of satellite communication between ships and the shore, where a satellite receiving station was temporarily installed at the Burnham site. A press report from the time states;

"An Anglo-American project in ship-to-shore communication via space satellite, which could provide instant contact between passengers and their homes, gets off the ground today when Atlantic Container Line's vessel *'Atlantic Causeway'* leaves Liverpool's Gladstone Terminal for the United States. Aboard *'Atlantic Causeway'* is a complex set of sophisticated equipment, supplied partly by the United States Coast Guard, partly by the Department of Electrical Engineering at Swansea University College, and partly by the GPO. This will be used to carry out a series of tests, sending a wide variety of messages to the shore radio station, at Burnham via a NASA satellite floating 24,000 miles up in space.

Satellite communications testing in 1971 at the Highbridge site.

"The main advantages of satellite communication are greatly improved clarity coupled with greater reliability" said Dr. Raymond Macario, senior lecturer in electrical engineering at Swansea University College. "At present, if you want to make a telephone call or send a message by teleprinter to a man aboard a ship at sea, you are subject to innumerable delays. The call has to be booked, the ship traced, and often the message is badly distorted by poor weather communications.

"If communication by satellite is proved to work there will be a vast improvement which could, for example, speed up personal contacts between ship and shipowner, and greatly assist marine rescue operations. Looking ahead 10 years, you could have 24 hour a day high-quality communication between businessmen and their offices, passengers and their homes, and crew and their families from ships anywhere in the world. "The equipment has already been tested extensively on shore. Now it remains to be seen if it will work perfectly under normal operating conditions at sea. Tests will be carried out for about 90 minutes a day, and will go on for four months". *'Atlantic Causeway'* was selected for the tests because she operates a fast regular service across the Atlantic, on a route well within the range of the NASA satellite. Also, since she returns home at fairly rapid intervals, items of equipment and personnel can be changed round as required. The equipment itself is completely automated and requires relatively little expertise to operate. At present, use of the satellite has to be hired from NASA, but if the operation proves a success, consideration will have to be given to the purchase of a new satellite for this specific purpose. This would probably cost in the region of about £10 million".

It does appear that these tests were indeed successful, as a further report in 1971 confirms;

"Ship-to-shore links through satellite with ships at sea, through the Post Office's ship-to-shore service have been proved to be technically feasible by recent tests. During the last few months, the Cunard-Brocklebank containership *'Atlantic Causeway'* played its part in test transmissions. The possibility of a commercial ship-shore service by satellite is now being undertaken by several other shipping nations. The tests, which began in the summer, were carried out by the Post Office, shipping and radio operating companies, suppliers of equipment, helped by the National Aeronautics and Space Administration of America (NASA) and the US Coast Guard. Good quality voice circuits were reliably maintained using relatively simple aerials and equipment over a VHF radio circuit from the Post Office Radio Station at Burnham-on-Sea. Somerset. A satellite was made available by NASA. There were also successful tests with transmissions of teleprinter copy, facsimile documents, and computer data. Among the organisations co-operating in the tests with equipment, staff and services were the Aveley Electric Ltd, British Aircraft Corporation, International Maritime Radio Committee (CIRM), Chamber of Shipping of the UK, Cunard Brocklebank Ltd. Department of Trade and Directorate of Telecommunications, Kelvin Hughes and Co Ltd, Marconi International Marine Co Ltd, The Marconi Company Ltd, Muirhead Ltd, Redifon Ltd, SE Laboratories Ltd, Standard Telephones and Cable, Ltd, Swansea University Department of Electrical Engineering, Swedish Ericsson Ltd, United States Coast Guard".

The end of an era came when the Post Office WTS (Wireless Telegraphy Section) issued the following statement in early 1971;

"The long range area communications scheme will end at 2400 GMT on the 31st July 1971. Details of subsequent arrangements for handling long range W/T traffic are contained in Notice to Ship Wireless Stations No. 2 of 1971. Copies of this notice have been distributed to Shipowners for forwarding to their vessels. Radio Officers are asked to make enquiries from their Owners if they have not received their copy of this Notice".

As a consequence of this action, the RN presence at the station ceased. The last RN Operator left the station during 1971, although Naval vessels continued to use the station for R/T calls on a regular basis, and the occasional ZBO was received by the civilian staff.

One of the last Naval operators at GKA, 1971.

The cessation of the area scheme was lamented by many seagoing Radio Officers. The July/August 1971 issue of 'Signal' (The Journal of the Radio and Electronic Officers' Union) featured the following tribute on the back page;

Farewell – Area Scheme: The few remnants remaining from the Second World War are rapidly disappearing, one of the latest to go is the 'Area Scheme' slipping quietly into the grave at midnight on the 31st July. Seemingly not worth any 'air-time' on the BBC Merchant Navy Programme, fortunately however Cape Naval Radio 'saved the day' and oozed the following MBMS in the pangs of death;

= MBMS =

FOR TWENTY-FIVE YEARS OR THEREABOUTS THE SAME OLD CALL SIGNS HAVE GONE OUT NOW ZSL HAS GOT THE RUB AND ZSJ HAS AN EXCLUSIVE CLUB. FOR ALL THE BOO BOOS WE HAVE MADE FOR THE SAKE OF AULD LANG SYNE? FORGIVE OUR TRESPASSES WE PLEAD, AS WE HAVE FORGIVEN THINE. SO RAISE A GLASS AND DRINK A TOAST IN SPIRITS, BEER OR WINE WITH US WHO LATELY WERE YOUR HOST TO DAYS OF AULD LANG SYNE =

CAPE NAVAL RADIO +

It is such a pity that these 'would be' poets have been hiding under their Morse key for twenty-five years, imagine what Portishead Radio could achieve in the literary field.

R. Dewar Twist, Radio Officer – Hudson Venture

At sea, 1st August 1971.

As a result of the cessation of the Area Scheme, vessels were required to contact Portishead directly. To assist in this, a special 'Pacific Watch' was set up specifically to listen for vessels calling from the Pacific Ocean, using directional receiving aerials.

Portishead Radio R/O Steve Elliston was at sea at the time and recalls;

"I was at sea from 1968 to 1975. In 1968 British ships, wherever they were in the world, used the 'Area Scheme' to receive and send traffic. For its day it was quite a good system though it could lead to delays as the traffic transited the Naval communications networks.

It was running for years and, as a result, all British ships needed for HF communications was a minimum fit, i.e. a 100w transmitter e.g. the Oceanspan and a receiver like the Atalanta. Foreign ships even at this time had, in the main, much better equipment e.g. 1.2 kW transmitters. Then the area scheme ended and British ships had to work direct i.e. back to GKA, from wherever they were.

Because of cheap tariffs, GKA was incredibly busy so there were a lot of ships calling and being QRY'd. As you know sometimes a QRY 30 could disappear in 20 minutes or so; or, more usually, 1 to 2 hours. In dodgy areas like the Gulf of Mexico or the Pacific, the chances are GKA would have faded out by the time your turn had arrived and the 45 minutes spent calling to get the QRY30 would have been wasted, never mind the 1 or 2 hours waiting out the QRY. Frustrating.

Coupled with that, some of the Portishead transmitter frequencies were swamped by more local transmitters, San Francisco/KFS on 8 MHz being a good example – though to their credit the PO/BT did bodge some of the transmitter frequencies to try and give them a bit more separation but a QSA5+ multi kW transmitter in California versus a QSA1 GKA transmitter on an Atlanta did make exchanging traffic more than a bit difficult! Of course, the excellent Rogaland/LGB or Lyngby/OXZ could be, and were, used, but that was just to clear outgoing traffic – incoming stuff was still in the carousel at GKA.

So after much pressure GKA started the Pacific Watch – different areas/sectors had different times, there was a schedule - one or two blokes searching specific calling frequencies with rhombics aimed accordingly I found it pretty useless. My transmitter was a Crusader or an ST1200 – meaty transmitters and well up to the job, but only occasionally was I successful in working back. It was often possible to work back normally and get QSA 3-4 whilst the Pacific Watch was N/R.

I eventually found what my problem was – my transmitter frequencies were off. Transmitters in those days were crystal controlled – when the scheme started Head Office duly forwarded me additional crystals. What they didn't do was check they had the correct holder capacity; consequently my crystals, although labelled correctly, had the extra few pF's enough to cause my transmitter frequency to be 1 kHz or so off. I bodged the crystals and after that the Pacific Watch was great!"

Another R/O was quoted as saying that "There is no doubt that the introduction of the Pacific scheme has greatly improved contact with Portishead Radio. Without it we had sometimes to call for hours. Now I personally have been getting GKG through one or two short calls most of the time. At present GKG works on one band at a time which sometimes happens to be useless in the area for which the service is meant. I do not blame GKG for that. The ionospheric conditions change unexpectedly. I would be much happier to see the 'Pacific Scheme' operate on more than one band simultaneously. This would certainly improve the contact".

Some seagoing R/Os found the Pacific Watch very much to their liking. Tim Strickland has fond memories:

"GKA was one of the best, if not the best, HF (long-distance) stations I used to work at sea. Contacting them by CW from the South Pacific was always a challenge, especially as the transmitter I was using (the ubiquitous Oceanspan) had an output power of about 90W. The professionalism of GKA's Radio Officers being able to pick out my weak signal in the mire of others never ceased to amaze me".

It was around this time that the Burnham Radio/GRL MF c.w. broadcasts on 1612 kHz ceased. Although popularly listened to by coastal vessels, it was decided that the short-range stations around the coast of the UK provided adequate coverage for weather transmissions.

In fact the Pacific Watch was described in a 1973 Post Office publication:

Ship Frequencies: 4186.5 6279.75 8373 12559.5 16746 22262.5 kHz.

Portishead Radio Frequencies in the GKG Family.

Sector Monitoring:

Panama: 1700-2100

Frisco: 2100-0100

Hawaii: 0100-0430

Wake: 0430-0900

Luzon: 0900-1300

Malacca: 0000-0800.

Due to increasing volumes of traffic following the withdrawal of the Area Scheme, a three-phase programme was undertaken to reconstruct the long range service. A paper from 1973 confirms the plan discussed earlier.

Phase 1 began in 1971 and provided additional transmitter and receiver capacity. To meet this requirement, 11 redundant transmitters at the Dorchester site (previously used for the fixed service) were brought into operation for the maritime service, and fitted with omni-directional aerial systems. Twelve new receiving positions were installed at Burnham in a prefabricated building erected adjacent to 'A' Wing and was to be known as 'D' Wing.

Phase 2 was a long-term transmitter provisioning programme. Phase 1 was very much an interim solution, so 12 transmitters at the Leafield fixed service station were utilised for the maritime radiotelephone and radio telex operation. Other requirements were fulfilled by use of transmitters at Ongar and Rugby. Replacement aerials were installed, being high-performance omni-directional and rotatable directional types (the latter being controlled by an operator at Burnham) suited for the maritime service.

Phase 3 was the reconstruction of the receiving station at Burnham, although consideration was given to transferring the whole operation to Somerton following the increase in housing developments close to the Burnham receiving site. After due consideration the move the Somerton never took place and the new building commenced construction in 1976.

17th June 1972 saw the start of the Observer Single-Handed Transatlantic Yacht Race, the first 'large' yachting event to use the R/T service at

Burnham. Full radiocommunication guidelines and propagation guides were published in association with the race organisers, something which was a regular occurrence for all subsequent yacht races up to the end of the 1990s.

Both double sideband and Single Sideband methods of R/T communication could be handled, although Portishead would reply using single-sideband full carrier (A3H) when replying. Details of coast station frequencies were also advised. Portishead would also broadcast dedicated weather information to all competing yachts on a daily basis.

However, this race can probably be regarded as the first major race to use Portishead Radio for all services, and thus started the excellent relationship the station has had with the leisure industry. Further details of future yacht races are described later in this book.

Again in the early 1970s, a 'trawler watch' was set up using the GKK transmitters. The role of the watch was to communicate with the vessel *'Miranda/GULL'* in sending meteorological reports on a regular basis as well as exchanging normal traffic.

'Miranda' was a Government-owned ship provided by the Department of Trade and Industry (DTI). She was operated by HM Coastguard under the direction of the Marine Division of the DTI and was managed by the Ellermans Wilson Line of Hull.

Her Master (Support Commander) and Medical Officer were appointed by DTI, and a Meteorological Officer was provided by the Met Office. Marconi Marine provided the Electronics Officer and Radio Officers. Ellermans Wilson Line provided the other officers and the crew, which included a fishing adviser (ex-Trawler Skipper) to assist the Support Commander in technical matters peculiar to trawlers.

'Miranda' was 'on station' in the Denmark Strait (between Iceland and Greenland) from the beginning of December until the end of April and crew reliefs were initially effected by air transport at Reykjavik and/or Isafjordur in Iceland. During the 'Cod Wars' reliefs were effected at Lerwick, Aberdeen or her home port of Hull.

Twice daily at 0940 GMT and 2140 GMT, *'Miranda'* conducted a 'round-up' at which all British trawlers in Icelandic fishing grounds reported their position by means of a lettered grid. On completion, the Met Officer gave the weather forecast and following this, any advice or information which the Support Commander wished to give was broadcast. Finally, trawlers were asked to pass in any requests for medical advice or attention and for technical services.

'Miranda' maintained a constant radio watch on 500 kHz (W/T), 2182 and 2226 kHz telephony and Channel 16 VHF for emergency calls for

advice or assistance. A constant HF Morse link was also maintained with Portishead Radio (GKK) with a dedicated operator where constant traffic of support nature and weather observations (OBS) were exchanged.

The service continued until 1980 when *'Miranda'* completed her last tour of duty.

Britain in the 1970s was strewn with industrial action and the station was no different; strikes, overtime bans and 'work to rules' took place on a few occasions during the decade as the Unions fought for pay rises and parity between R/O grades at the station.

There was a staff overtime ban from March 19th until March 25th, 1967, and in 1968 there was an unofficial strike between 1200-1400 GMT daily, between February 18th and February 23rd inclusive. This was then followed by another unofficial strike between August 2nd (2200 GMT) until August 3rd (2200 GMT) in 1970.

R/O Bryan Richards recalls:

"The strike in 1970 was for RO1s. I was an RO2 at the time and we were expected by the UCW to come out in sympathy. RO2s got nothing out of it. We were all civil servants under the GPO. All RO1s at the time were under parity of conditions under the Whitley Council which covered Diplomatic Wireless Service, GCHQ, RAF, Interpol, etc. The GPO, being somewhat commercial, was always trying to undermine the other groups' advances in conditions and pay. Hence on one circumstance non-W/T work was downgraded, hence the introduction of RO2s".

Bryan goes on to relate:

"The ending of the two-grade Radio Operator system at the coast stations started with the withdrawal of the RN at the end of the Area Scheme in 1971. The first batch of RO2 promotion to RO1 was 13 at GKA – I was one of them. The agreement was then a period of promotion and radio college for 2nd Class ticket R/O's. I was RO2 rep at the time and got a lot of stick".

Newspaper reports on the 3rd August 1970 stated;

"The *'Queen Elizabeth 2'* was among ships affected today by the cut in ship-to-shore contact caused by an unofficial 24-hour strike by Post Office Radio Operators. A spokesman for Cunard said the *'QE2'* was out of touch with the U.K. except for distress and safety calls. Earlier a spokesman for P & O, another leading passenger line, described the stoppage as serious. One hundred and thirty workers were cut at the Post Office long-distance station at Portishead, Somerset, and others stopped work at Ilfracombe".

On the 5th October 1970, further action took place as national media reported;

"Wireless operators at three Post Office marine radio stations were still on strike today. Men at Portishead, Somerset, high-frequency station who handle traffic from ships all over the world, and the Ilfracombe, Devon and North Foreland, Kent stations, which cover coastal traffic, went on strike yesterday. The operators claim that their pay should be linked with that of the Merchant Navy rather than that of Civil Service operators".

Portishead R/O Mike Howley recalls some of the background to the industrial action of the 1970s;

"For quite a few years prior to 1970, the R/Os' salary was tied to that of the salary of the Composite Signals Organisation (GCHQ) Radio Officers. Years earlier a high court judge had ruled on a previous salary dispute between differences between R/Os' salaries in the CSO and the WTS (as it was then). The judge had said that "he was not going to have the tail wagging the dog". (There were 2000 CSO R/Os and less than 150 WTS R/Os at that time.) The settlement was therefore fixed that the WTS R/Os would have the same salary as CSO R/Os.

However in 1968, the Civil Service Union threatened strike action, claiming that seagoing R/Os were paid a great deal more than CSO R/Os. An arbitration tribunal was held in London (to which only persons who had signed up to the Official Secrets Act up to the level of being able to deal with documents classified SECRET were admitted. At this time the very existence of GCHQ was a state secret).

I was called as a witness in that arbitration tribunal because I had fairly recently been a seagoing R/O rather than a GCHQ R/O.

As a result of this arbitration tribunal, the salary link between the CSO R/Os and the WTS R/Os was broken. A year or two later the UPW claimed that WTS R/Os salaries should be linked to that of seagoing R/Os. The UPW (Union of Post Office Workers) was claiming that WTS R/Os' salaries should be doubled!

From the Post Office's point of view, the problem was not so much the total wage bill for R/Os (there were not many of us remember), it was more that the differential between R/Os and other PO grades would cause immediate wage claims of many other grades. The PO got around this by offering the double-time on Sundays, the time and a half on Saturdays and various other shift-related enhancements.

In some ways this pay settlement did work. When BT was separated from the PO and privatised, the Annual Accounts presented to shareholders had to show all employees who earned more than a certain amount. For some years the Annual BT Accounts presented at AGMs showed the salaries of the top executives in BT HQ plus those of two or three R/Os up at GKR who had earned director-level salaries! (This was the result of a lot of extra duty of course).

As regards the management strikes, I can remember nothing about the actual claims. I do remember one amusing incident, however. The morning after one of the R/Os' strikes Assistant Superintendent Ken Gill was standing at the entrance and greeted Jim Sheldon with "Good morning striker Sheldon".

A year later when Ken Gill had been on a management strike, Jim Sheldon deliberately positioned himself by the front door and as Ken Gill came in, he greeted him with "Good morning striker Gill".

Ken Gill just scowled at him, which amused Jim Sheldon. He knew his barb had struck home!"

On 21st May 1975, the national press reported;

"Portishead Radio Station at Burnham, Somerset, Britain's only long range link with ships throughout the world, was closed last night for the second night running by strike of 150 radio operators. The operators, members of the Union of Post Office Workers, are understood have rejected a pay rise offered them earlier yesterday".

Union representative Gerry Knott addresses the staff in the car park.

A further report advised;

"Industrial action at radio stations, including Stonehaven and Portpatrick, is affecting Ship-to-Shore communications, the Post Office said yesterday. Operators at seven radio stations stopped work yesterday afternoon in support of a pay claim. A statement said commercial telegraph and radio telephone services had been stopped but all distress services were being fully manned. Coast stations were continuing to broadcast navigation warnings to shipping. Other stations hit by the dispute: Portishead Radio (long range) and coast stations at North Foreland, Land's End, Ilfracombe, and Anglesey. Although radio operators' pay was not due for review until June 1, the Post Office were preparing to make a pay offer within the next 24 hours, said the statement".

In March 1975 the Post Office introduced a pilot scheme for selective calling in the HF bands using the Sequential Single Frequency Code {SSFC} system. This pilot scheme was intended to last for five years, but in the event it proved expensive to operate and the number of ships using the service was negligible. It was closed down on 31st March 1980 but continued to be used for MF and VHF coast stations.

Alan Hewitt was a Naval R/O at the station, and he recalls;

"At that time there were 5 RN operators in total and our watches were split into 0800-1300, 1300-2200 and 2200-0800 (nights). There was only one RN operator per shift and we only worked on 8 MHz c.w. ship-shore.

I was scheduled to work the 0800-1300 shift and en route to the radio station in my Morris 1000 I was a little apprehensive at the thought of crossing a picket line but on entering the main gate saw just 2 or 3 of the staff there and they waved me through. I believe the strike did not start until 0800 so that would account for the low numbers at the main gate. By 0900 quite a crowd had gathered but I do not recall any shouting or waving of placards and it seemed to be all very low key and peaceful.

The RN operators had each received a memo from our boss (Lt G Lewes RN) several days before the strike telling us to report for duty as normal but only work RN warships or vessels in distress. On no account were we to work merchant vessels. Despite the fact that a message had repeatedly been sent on the Broadcast on all frequencies informing ships of the strike date and not to call GKA, several merchant ship operators seemed to have ignored the advice and quite a few (especially foreign-owned ships) were calling on 8 Mhz. I ignored them all and worked just one warship during my watch. It was easy to know if it was a warship as the RN used the 'Z' code rather than the 'Q' code. (ZBO from RN ships was the equivalent to QTC in the MN). Also, there was a call sign book kept next to the Radio Overseer that we could refer to. There was one other staff member in the wing in case of an emergency and also an Overseer sat in his usual position. It was a boring shift.

I returned at 2200 the same day for my night shift and all was back to normal".

September 30[th,] 1972 saw the end of the GTZZ 'Press' service, as advised in the November 1972 'Wireless World';

"September 30th marked the end of an era in shore-to-ship news transmissions, with the closedown of Shipress (GTZZ) by Wireless Press Ltd, London. Reductions in many passenger fleets led to the decision to sign off an automatic, twice-nightly Morse service to ships at sea. The final message read:

WIRELESS PRESS SENDS GREETINGS TO ALL SUBSCRIBERS AND READERS ON THE OCCASION OF ITS FINAL SHIPRESS TRANSMISSION. BON VOYAGE.

First experiments in regular transmissions of news telegrams to ships involved Browhead, Ireland, wireless station and the Cunarder *'Etruria'* on February 7th, 1903. Later that year the Cunarder *'Lucania'*, ex- Liverpool, assisted with the inauguration of a daily news service via Poldhu, Cornwall and Glace Bay, Newfoundland. The Wireless Press Ltd, which originally published Wireless World, was owned by the Marconi Company until its acquisition by Associated Iliffe Press Ltd, in 1924. Post Office stations involved from 1926 were Leafield, Oxfordshire, Rugby and latterly, Portishead. Principal news agencies contracted over the years were Reuter and Press Association and since 1967 the London bureau of United Press International prepared editorial copy for the service which was received by Radio Officers aboard troopships, ocean-going yachts and cargo ships in addition to British, Italian, Dutch, Greek, Norwegian, Polish, French and German liners".

Most owners of passenger ships subscribed to this press service. Subscription press from the UK was sent from Portishead Radio and addressed to the collective call sign GTZZ. Addressing the broadcast to a collective call sign made the content private rather than broadcast, and it was thus an offence for R/O's on non-subscribing ships, or indeed anyone else, to copy the press. GTZZ broadcasts came in two, three-quarter-hour long sessions, the first came at 2148 GMT, followed by the second at 0100 GMT. The content of these messages enabled the production of a daily ships' newspaper of some six pages of typescript. This included some selected horse racing results and a whole page of stock exchange quotations.

The broadcasts were sent from the Broadcast Wing (B Wing) at Portishead and were transmitted using 5-unit tape at a set speed of 20 words per minute. However, it was not uncommon for the speed to run in excess of 25 words per minute, depending on the R/O in charge of the transmission.

In order to be confirmed 'in post' at Portishead Radio, all newly-employed Radio Officers had to undertake a probationary period of one year, during which they had to pass numerous examinations. The official guidelines from 1974 show how involved the process was;

TRIAL PERIODS – RADIO OPERATORS

1. TRIAL PERIOD

1.1. All Radio Operators are required to serve a trial period of one year from their date of entry. Second Class Certificate holders are required to serve a further 12 months trial period from the date of which they obtain a First Class or General certificate.

1.2. Post Office rules governing trial periods apply to all Radio Operators.

1.3. During trial periods radio operators are required to pass the tests detailed below.

1.3.1 Tests for Radio Operators holding a Second Class Certificate and for First Class Certificate holders who fail to become fully qualified. During the trial period these Radio Operators are required to pass the following manipulative tests;

Morse: Sending and receiving plain language and code groups at the same standards as in 1.1 (training). In addition, sending and receiving figures at 12 groups a minute for one and a half minutes; sending and receiving eight groups of accented letters in one minute. A maximum of two errors is allowed in each figures and accented letters test but there must be no uncorrected errors in the sending tests.

Teleprinter Operation: Sending 25 messages in 30 minutes with no uncorrected errors and not more than ten corrected errors. The messages are of 17 words average length including the preamble.

1.3.2 Tests for Radio Operators holding First Class or General Certificates. During the trial period these radio operators will be required to pass the following tests in order to become fully qualified;

Language: Those who do not already hold the elementary certificate in French issued by the Royal Society of Arts or certificates of an equivalent standard are required to pass the Departmental test in this language. Those who hold certificates issued by a recognised examining body should submit them to ETE/MRSD (External Telecommunications Executive/Maritime Radio Services Department).

Manipulative tests: The candidate is required to:

Send and receive at the rate of 27 words a minute for five consecutive minutes without error or correction. The test is conducted with double headgear telephone receivers with weak signals;

Use a standard telegraphy typewriter to type messages being received in Morse at the rate of 18 messages in 30 minutes. The average length of a message is 12 words, counting the address and text only; a 20-second space is allowed between messages for checking the number of words and entering the received particulars. Five corrected errors are allowed but there must be no uncorrected errors. Messages forms are rolled and perforated and the test is conducted with double headgear telephone receivers;

Send 25 messages by teleprinter in 30 minutes. The length of a message is 17 words including the preamble. Ten corrected errors are allowed but there must be no uncorrected errors.

Practical test: The purpose of the examination is to ensure that an officer has acquired a sound working knowledge of all the apparatus at his station. Officers are given every facility to prepare for the test; at each station, the necessary diagrams and technical literature are available to the staff. The candidate is questioned on the working, adjustment, and maintenance of the items in the following schedule and, in addition, may be required to demonstrate soldering techniques and the use of testing instruments.

Landline equipment – Teleprinters, Keyboard perforators, Wheatstone transmitters and associated control apparatus; telephone link equipment and associated apparatus.

Radio equipment – Valves; motors; generators; power supplies; receivers; transmitters; serials; control panels.

Internal Combustion engines – Starting arrangements; lubricating systems; fuel supply; cooling systems; governors.

This test continued (with a few changes) for all new entrants until recruitment ceased in the early 1980s. A German-language test was once part of the examination, and when electronic Morse keys became available, a separate test was required on these to ensure the operator could use one competently.

On February 1st, 1978, the station became a full member of the Automated Mutual-assistance Vessel Rescue (AMVER) system. This allowed vessels to use the station to forward regular ship information, position and voyage reports to the AMVER centre in New York, which would be held on a database for instant retrieval and transmission to rescue agencies world-wide. Such messages would be free of charge to the vessel, the extra cost of Post Office transmissions being paid for by the Department of Trade. These reports were extremely popular (and sometimes very detailed), with vessels providing details routes and waypoints for their current voyage.

Interestingly the AMVER system remains operational today, with reports being tendered by vessels using satellite-based systems.

Radio Officer Jim Byrne accepting the AMVER flag in 1979.

One interesting incident which occurred in July 1978, relates to the vessel *'St.John III/6ZKX'*. The R/O on board, Charles Marshall, tells the story, which is worth relating in full;

"We waited for 5 days for sailing orders from Texaco to the Escravos River or Bonny, Nigeria to load crude oil for Tema, Ghana. After 2 or 3 days delay we weighed anchor and proceeded on the owner's instruction. Somehow, and why I will never know, the contract was cancelled by Texaco and the ship was instructed to proceed to Abidjan and anchor in Lake Ebrie.

The ship arrived on July 13th. While there waiting for a cargo we proceeded to carry out much-needed repairs and 'Dig Tanks'. As *'St. John'* was a little past her prime, much work still needed to be done. Time passed quickly (i.e. the previous R/O had taken the Direction Finding equipment to bits, even the goniometer, so I was busy trying to put it back into working condition. Even the auto alarm equipment was broken.

On the evening of the 14[th,] a most irate Ship Chandler appeared on board with his mixed entourage demanding payment of long outstanding bills for supplies and victuals, etc. He informed our illustrious Captain of this in a threatening manner, who in turn informed him politely to take it up with the civil authorities. At this stage the Ship Chandler who wanted his money lost his cool; he took some matches from a box and struck them very close to an open cover of a venting tank (as we were still not gas-free this was considered a little hairy, to say the least). At this point, the Captain told him to put the matches out or we would all live to regret the situation. The Ship Chandler refused to do this, so the Captain took the initiative, dashed to his accommodation and retrieved the ship's pistol. The Ship Chandler was not impressed by this and proceeded to strike more matches. The Captain, seeing this, and noting also that this chap was getting even closer to the venting tank, fired a single shot into the jungle which was about a mile away! The Ship Chandler panicked, dropped his lighted matches and ran, screaming that he would return with the police the following day. Anyway, after all this excitement the night passed peacefully.

In the early morning, the vessel owner's representative, who had flown out from the UK the previous day, arrived on board. The Captain advised him what had gone on, the representative turned a funny colour and then disappeared down into the depths of the midships pump room until about 1100 hours. Coincidentally it was at this time that the local Constabulary, along with the Ship Chandler arrived. After a few raised African voices they departed some 30 minutes later, the ship being placed under arrest. The Captain, the Chief Steward, the Chief Engineer, and the Owners Representative (all British subjects) were walking down the gangway also under arrest, and as I am leaning over the ship's rail trying my best to look serious, the Representative shouts up "Sparks – call in the cavalry!" I think to myself 'who are the cavalry?', so after consultation with the Mate, I warmed up the main transmitter and call Portishead Radio on R/T. Now, they think that it's that chap on '6ZKX' trying for a normal call and try to give me turn number 5 or something. I managed to break in and advise the Portishead operator that I did have a very urgent situation. He was most helpful and immediately opened another channel for me. After explaining the situation they put me through to the owner's headquarters in London, who in turn informed the Foreign Office, P and I club, etc.

By this time, the 'party' from the ship was ashore at the Police Station, where obviously the appropriate statements from the different parties were being taken in Pidgin English and schoolboy French. The party was then put in a waiting room. In the meantime, the Foreign Office had sprung into action (10 out of 10 for them!), and the British Embassy representative appeared at the Police Station within 2-3 hours. The Embassy representative interviewed the Captain and found out what had happened, and they were all then escorted into the Chief of Police's office where the 'much aggrieved' Ship Chandler was waiting. The British Embassy representative then explained to the Chief of Police in what I was told sounded like the most eloquent French ever spoken what had happened from the ship's point of view. Upon completion of his explanation the Chief of Police looked at the Ship Chandler gravely, and the Captain benignly. He returned the Captain's confiscated gun to him and the 'Ship's party' was escorted from the Police Station back to the ship; we never saw the Ship Chandler again.

This incident I know caused much levity at Portishead Radio. How do I know this? Well, the 'smile' in the voice of the 'old man' at Portishead to whom I was speaking, and the giggling in the background (I know this was not from my end!) from then and thereafter caused me to realise that the adventures of *'St. John III/6ZKX'* were followed as closely as those at the Crossroads Motel and Coronation Street!

Thanks, Portishead, for all your help, understanding and absolute professional quality of your work over the years while I was at sea. YOU WERE THE BEST".

The Dorchester transmitting site closed down in July 1979 as part of the Post Office's rationalisation programme, following the closure of the Portishead site on December 31^{st,} 1978. This left Ongar, Leafield, and Rugby as the only remaining transmitter sites used for the maritime long range radio service.

The closure of the Portishead site was in many ways the end of an era. The station had been a significant landmark in the area for over 50 years and had provided an excellent service. However, the equipment was becoming more difficult to maintain and existing transmitter sites could perform the same tasks. The Post Office therefore decided to rationalise the numerous radio transmitter sites, resulting in the closure of this historic station. However, the station continued to be known as 'Portishead Radio' until the final closure in 2000.

The site is now used as the headquarters of the Avon and Somerset Police.

The felling of one of the Portishead masts, 1979.

In April 1979, a discussion document entitled 'Maritime Services Review of the Future' was published. This extensive document covered all parts of the service, from VHF to HF, but the Long Range HF service section is of most interest;

"The terrestrial Long Range Services are seen as having a diminishing but still very significant part to play at least until the end of the century despite competition from the satellite service. Several major programmes are already in hand to ensure our ability to meet these requirements.

1. **HF Rationalisation Programme.**

Under the HF Rationalisation Plan prepared in 1971, a total of 60 HF transmitters have so far been transferred to the maritime service from the fixed services and appropriate aerial systems provided. These transmitters are located at the main HF stations and remotely controlled from Burnham. Whilst there have been some shortcomings with certain design features of the remote control and transmitter drive arrangements that are being corrected, there is little doubt that we probably have the most powerful maritime HF transmitting capability in the world. This provision should meet foreseeable demands. Replacement of time expired equipment may come up during the late 1980s and will be considered on its merits at the time.

2. **Burnham Reconstruction Programme**

The second major programme concerns the reconstruction of Burnham receiving station. The need to replace the existing Burnham station was recognised in the late 1960s. The buildings were old and substandard, and inadequate to accommodate the expansion of the services. Moreover, the aerial systems were increasingly threatened by radio noise form the residential development that was taking place around our very restricted site. As a result, it was proposed to rebuild the station, but at Somerton Radio station where there was already an excellent receiving aerial system, not subject to the shortcomings of Burnham that could be made available and modified for use by the maritime services".

As is well-known the proposal did not materialise. Eventually (1974) with the availability of suitable remotely controlled equipment it was possible to develop an alternative plan which made use of the Somerton facilities but did not require the transfer of operating staff from Burnham. It is emphasised that it only the arrival on the market of the remotely controlled equipment that enabled the impasse to be broken. Bearing in mind the close proximity of residential development to the existing site and its potential for degrading the performance of a radio receiving station it was considered too risky an investment (£1.5-2m) to have replaced the radio receiving facilities at Burnham.

In the event, the reconstruction programme was authorised (1976) on the basis described above with the aim of providing operational and welfare accommodation and facilities adequate to meet the requirements during the next 20 or so years.

"The system will comprise some 50 Morse operating positions, up to 18 radiotelephone circuits and up to 14 radioteleprinter circuits. To meet the changing demands of the future the building has been designed for maximum flexibility and the floor of the operating room is of the computer room type that will permit rearrangement of operational layouts with a minimum of constructional upheaval.

The opportunity has also been taken to adopt computerised message distribution techniques for the Morse telegram service that will avoid the circulation of hard copy and the use of conveyor belts. Under the latest proposals 'to-ship' telegrams will continue to be accepted at Burnham but 'from-ship' messages will be sent direct from Burnham to the TRC (Telegram Retransmission Centre) for onward disposal to the addressees. The facilities of the TRC will also be used to hold in store the ships' index file for Burnham and to prepare an accounting tape for off-line processing".

The summer 1979 issue of the staff magazine 'Hello World' carried an article on female Radio Officers, in which Heather Anderson was quoted as being the first female to be employed in the service. This prompted a reply from the Station Manager (Don Mulholland) who was pleased to put the record straight;

"Heather is the fifth in line of female Radio Officers employed by the Post Office, and their starting dates were;

Miss Lynn Burdon	7.6.76
Miss Odette Townsend	6.9.76
Miss Jennifer Evans	4.4.78
Miss Kay Griffiths	7.8.78
Miss Heather Anderson	6.11.78

Miss Anderson was the first to serve at a short-range coast radio station, although she is now serving at the long range station (Portishead)".

Besides those listed above, three further female R/Os were employed at Portishead, who joined in the early 1980s; Pierina Del Pizzo, Zoe Newton, and Cheryl Cottier, making a total of eight.

One duty that appealed to staff of a more technical nature was the 'Maintenance Officer' (MO). This was a panel of 4 R/Os who rotated duties on a weekly basis ensuring the telex machines, aerial amplifiers, typewriters, and other basic electrical equipment were maintained and working satisfactorily. Liaison with the local engineers took place on a daily basis and ensured all the equipment at the station remained in good order. However, the role ended in the 1980s when a new Chief Engineer was appointed and decreed that the MO staff were not suitably qualified to maintain any electrical equipment. The Panel resigned 'en masse' leaving the engineers to cover the more mundane duties, although a couple of volunteers were eventually found to cover the basic maintenance work.

It is interesting to look back at the W/T procedures in use for calling and working Portishead at the time. Without the Area Scheme to use, vessels wishing to use the station had to call directly to obtain or send traffic. The official procedures as described in the 'Guidance for Radio Officers' which was periodically issued by the Post Office (and subsequently BT) state;

"The Radio Officer should read the emissions on the GKB frequency before calling. If the signal DE GKB QRL is received or if he hears a QRY being given to another ship he should stand by. When the emission changes to a tuning signal followed by a channel indicator it is time to call. When the tuning signal changes it will be in the form DE GKB 3 (i.e. tuning signal plus an indication of the channel being watched). If you do not have a facility to transmit in that channel then you have to wait.

Assuming that you are able to transmit, your call should be short and in accordance with the regulations, viz.:

GKB (not more than twice)

De (your call sign)

QSS (followed by the last three figures of the working frequency)

K

If there is no reply, the call may be repeated at intervals of not less than one minute for a period not exceeding five minutes, and shall not be renewed after an interval of ten minutes".

Despite this, many vessels tended to send GKB persistently hoping that the station would ask for their call sign. However, this practice was much frowned upon, and the guide was clear to emphasise; "The Radio Officer guarding the calling channel will not answer ships who make long calls or engage in the practice which omits the ship's call sign and seeks to elicit DE from the coast station".

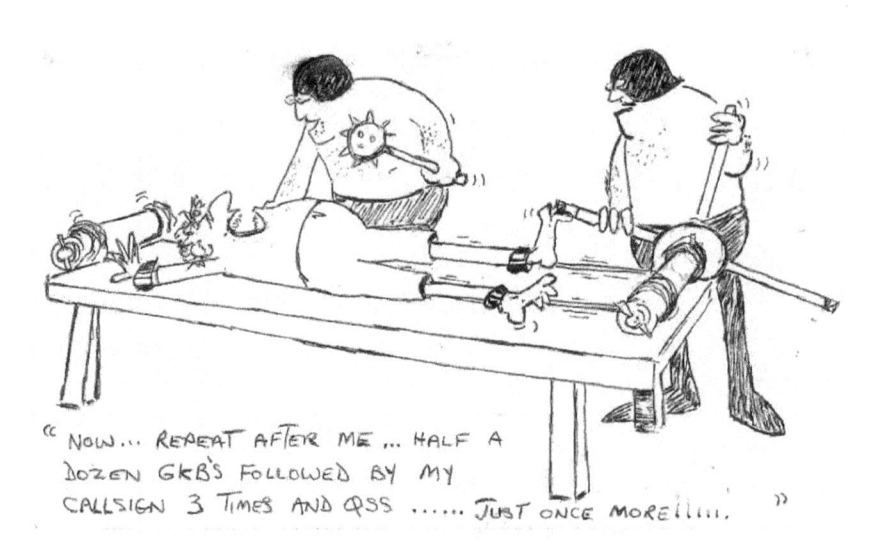

" Now... repeat after me ... half a dozen GKB's followed by my callsign 3 times and QSS ...... just once more!!!!! "

After the call was acknowledged, the Search Point operator at Portishead would advise the ship to monitor GKC or GKD and assign a QRY (turn) number. At peak periods this could be well over 50, by which time radio conditions may well have changed. The Search Point would also note the ship's call sign and aerial direction to assist in locating the vessel when being called by GKC or GKD. Any priority indicators from the ship (OBS messages or urgent telegrams) would bypass the QRY system and be handled expeditiously.

After obtaining a QRY or being told 'UP' by GKB, vessels would immediately retune their receiver to either GKC or GKD and change their transmitter to the working frequency given to GKB. As soon as the 'UP' signal was given by GKC or GKD, vessels would call on the working frequency to establish a workable connection. From-ship traffic would then be sent.

Should Portishead hold any traffic for the vessel, this would be transmitted on a clear working channel in the GKG-GKM series (GKF was also available on 12 MHz only).

Vessels in port could arrange schedules where any to-ship traffic would be broadcast 'blind' at a pre-arranged time and on a pre-arranged frequency (normally GKG). The messages would be transmitted twice before being held pending acknowledgement by the vessel on leaving port. This was known as the 'Point 43' duty at Portishead.

'A' Wing W/T operating area, late 1970s. Transmitter selection board visible on the wall. Racal RA1217 receivers installed on each console.

Broadcasts to ships took place on the GKA series of transmitters as shown in the table below:

Time (GMT)	Broadcast
Every hour on the hour	Traffic Lists
Every 4 hours from 0000	Collective Call sign messages
0930 1130 2130	Weather bulletins for the North Atlantic
0730 1330 1730	NAVAREA I broadcasts
0130 0530 0730 0930 1130 1330 1730 2130	Storm Warnings
0800 2000 (Sundays only)	Optimum Frequency (OTF) Predictions

Such broadcasts continued until the station closed in 2000, although the GKB transmitters replaced GKA in the final years.

The duties of staff were divided into morning, afternoon, day, evening and night shifts. Mornings tended to be from 0800-1300 or 0800-1340, afternoons 1300-1800 or 1300-1900, and evenings 1700-2300 or 1800-2300. Day shifts from 0800-1700 or 0900-1700, and there were two long afternoon/evening shifts from 1400-2200 or 1500-2300. On occasions some variation of the above took place. Nights were always 2300-0800.

Staff could choose if they could work a 5-day or 6-day week. Obviously, the same hours were worked, but the 5-day week had longer shifts (but 2 days off per week).

Within working periods, each shift was divided into hourly periods, with staff shifting between jobs on the hour. Some jobs such as R/T or Radiotelex lasted all or most of the shift, whilst others had a selection of jobs. Staff were free to exchange duties if required. Some examples of duties are listed below:

LL	Landline
R	Routeing
B	Bureau
P	Phonograms
SVC	Service Desk
TFC	Traffic List
E	Editing (to-ship)
RTT	Radiotelex
R/T	Radio Telephony
M	Monitoring
WTC	W/T Coordinator
CCN	Circulation

Specific HF W/T Search Points were listed as indicated below, as was the 'Point 43' blind broadcast position. If no designated duty was indicated, then normal HF W/T duties would be performed.

Most of the time was spent on HF W/T working, but designated Search Point working as allocated as follows:

Point 56 – 12 MHz

Point 54 – 16 MHz

Point 52 – 22 MHz

Point 42 – 4/6/8 MHz

Occasionally Point 44 was allocated if traffic on 4/6/8 MHz became busy to act as another Search Point.

This was probably one of the busiest jobs in the station, with staff being listed on a point for one hour only (although a few R/Os enjoyed the challenge of Search Point working and often stayed on for an extra hour). It was not uncommon for over 60 vessels to be given a turn number (QRY) in an hour, a rate of over 1 ship handled per minute.

"THIS IS YOUR LAST CHANCE GKB..... IF YOU DON'T ANSWER THIS TIME... I'M CALLING DAN...!!"

Each Search Point had different coloured chits to indicate the band of use. These would be completed with details of the time of calling, the calling vessel's call sign, aerial direction, working frequency and working transmitter (GKC/GKD). A switch would add 1 to the displayed QRY (turn number) and this would be passed to the ship. There was also space for any priority working (OBS, Urgent or État Priorité telegrams).

The colours of the chits were;

22 MHz – Red

16 MHz – Blue

12 MHz – Yellow

8 MHz – Green

6 MHz – Pink

4 MHz – Grey

These chits would be collected by messenger and passed to the circulation duty who would interrogate the 'carousel' where any to-ship messages were held in call sign order. Any messages were pinned to the chit and then passed to the WTC (Wireless Telegraphy Coordinator or Control) officer, who would then store them in time order, ready to pass to the next free working position. The QRY would also be decreased by 1 each time a ship was passed to a working position, which would also be displayed on the Search Point. The job of the circulation duty was to attach any to-ship messages to the incoming chit from the Search Point, and also to insert any incoming messages in the correct section of the carousel ready for the call sign to be added to the next traffic list.

Each working position was linked to the WTC by intercom and two indicator lights – a red one, indicating he was 'free' and a white one indicating he wishes to speak to the WTC officer. As soon as a working point became free he would be passed details of the next ship on turn, and any to-ship traffic passed to him by messenger.

Other jobs performed by staff at the time included 'Routeing' where from-ship telegrams would have their telegraphic addresses checked and telex number appended ready for landline delivery; 'Monitor' where an officer would check all incoming telex machines, respond to live queries by telex, and tear off all to-ship telegrams and pass to the editing position; 'Editor' where all to-ship telegrams would have their words counted and preamble added; and 'Landline' where from-ship messages were sent by telex to their destination. There were, of course, other duties to perform but these were normally of an administrative nature, such as dealing with service messages and checking for outdated or 'time expired' traffic.

The traffic list compiler sat on the other side of the traffic 'carousel' and it was always entertaining to see the carousel spinning round at high speed as both officers took traffic in and out of the unit every few seconds. At Christmas, there were often two officers on traffic lists and circulation to handle the amount of work – and one practical joker once attached a plastic 'severed hand' to the carousel to show how dangerous the job could be....

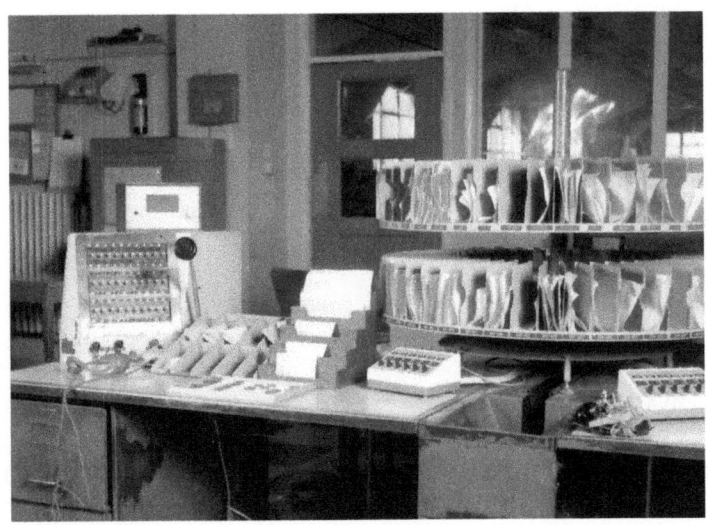

WTC position showing the traffic 'carousel'.

One of the most unpopular jobs was the 'Phono' duty where staff would telephone messages received after office hours to shipping companies throughout the UK. These were followed by the telexing of confirmatory copies during the evening and night shifts. Whilst most companies were happy to receive such telegrams over the telephone, many were happy just to see them on their telex machine in the office the following day. "Put it on the telex" was a common response from many companies, although some urgent messages were gratefully received.

Roger Marshall recalls one episode with the confirmatory copies duty:

"Dreaded confirmatory copies. I remember one night shift there were loads of them and the overseer wouldn't have a draw until they were gone. The R/O got sick of this, put them in his inside pocket and the pleased overseer finally gave a draw. It wasn't until later that day when said R/O remembered he had them in his pocket and on his return to duty quietly replaced them in the confirmatory box much to the dismay of the current overseer who was sure they weren't there during his previous check. Not sure if they were then telexed with 'delayed due error of service' or not".

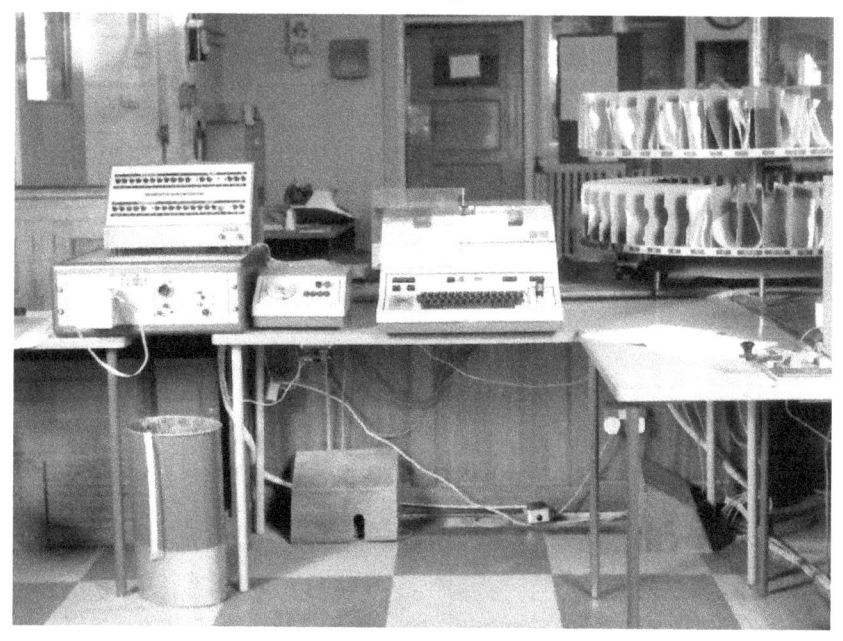

Traffic List position and the traffic 'carousel'.

Another role of this duty was to receive dictated messages (normally from families of ship's crews), which sometimes caused some amusement. I recall trying to transcribe a message from a lady whose strong Geordie accent was proving virtually impossible to understand. Luckily there was a colleague on duty from that part of the world who was able to translate for me.

Some amusing comments about the Portishead service at the time came from an anonymous Radio Officer who took great pains to compliment the station;

"On my last ship, a Norwegian flag vessel, I was directed as part of the charter agreement to work Portishead Radio. Incredibly enough, I had not much to do with Portishead for over five years. After much nervous reading and re-reading of ALRS Volume 1, I started up my 'Crusader' to send a telegram. It all went very well, and I found that the multi-transmitter system worked very well despite having to shift around a bit to receive messages. However, I did feel a bit sorry for blokes who might be working GKB without a digital read-our main receiver. I can hear enraged howls when I write this, but what was noticeable with GKB, for me anyway, after an absence of five years, was the politeness, patience, and skill of the operators. It can't be easy receiving and sending messages with QRY32, and half the ships involved unable to read Morse sent at 20 wpm. I had to laugh one Sunday morning when the bloke on GKC, after explaining for twenty minutes at 20, 15 and finally 10 words a minute to a Liberian that he should shift to GKH to get his message, sent a long series of dots and then politely asked the Liberian to call him back in five years".

Staffing figures from March 1979 show that the total number of R/Os employed at Portishead was recorded as 230, plus 5 in training, 9 overseers, 1 Head of Department and 4 Radio Assistant Superintendents. In addition, 4 telephonists, 6 handymen and 3 doorkeepers were employed. The total number of R/Os across the whole service (including coast stations) was 404, an increase of 52 over the previous year.

All contacts with the station would be kept in a log book, maintained on each working point. Staff would need to 'log on' at the start of the duty, and maintain details of ships worked, number of messages sent and received, and any items of note. The start and end time of each contact would also need to be logged, as well as times of 'grace reliefs'. If a ship was 'failed' for any reason, this also had to be logged in case of any future investigation. Examples of 'failure' would be severe fading (QSB), severe interference (QRM), high static levels (QRN) or simply the vessel did not respond to calls.

If the text of a received message had the potential to result in a possible safety issue (cargo shifting, equipment failure, the taking of water, etc.), the R/O had the option of asking the vessel's Master if he would allow details to be passed to Lloyds of London for insurance purposes. A 'May We Report Lloyds' book was held in the Control Room where such requests were logged together with the Master's response.

Mention would also be made of any suspect words which may have been received due to the poor quality of the Morse by the shipboard R/O. Such messages would be appended 'Sender's Risk' and duly logged. A book was held in the Overseer's office of vessels whose R/Os were deemed to be substandard and their shipping or radio companies duly advised.

The R/T log would also be used to compile the list of vessels queued up at a time, together with the number of calls made by each vessel. Details of the call itself (date, time, destination and duration) would be recorded on a chit and collected by the R/T Coordinator ready for passing to the accounts room for charging purposes.

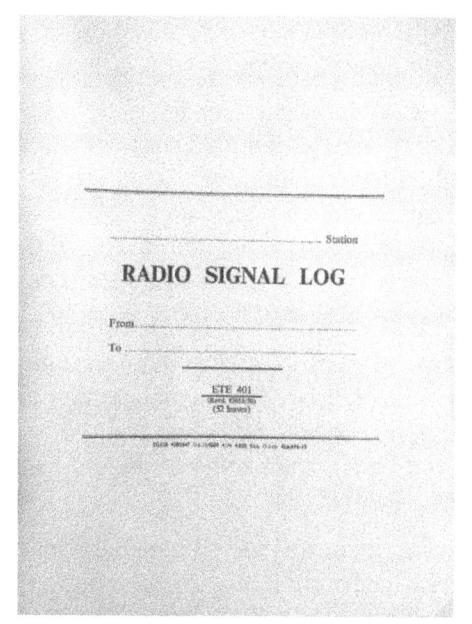

W/T and R/T log books from the late 1970s.

The late 1970s and early 1980s would see the staffing peak before the inevitable decline in terrestrial radio communications took hold, thanks mainly due to the increased use of the rapidly-developing satellite services and message handling technology for the maritime market. However, the station was well-prepared for the computer age, as the completion of the new operations building would confirm.

CHAPTER SIX

THE 1980S

The decade got off to an exciting start when a clerical error added a £523 'bonus' to the pay packets of staff in January 1980. Sadly the error was noticed and repayments were taken over the following two months.

Recruitment continued unabated as the new decade dawned. Two training schools were established, one for experienced seagoing Radio Officers and the other for newly-qualified staff without sea experience. The former course lasted for three months, with a high level of 'double-banking' with experienced staff members; the latter lasted for six months, with familiarisation of commercial working and touch-typing being high on the curriculum. Again, many of the inexperienced trainees spent a lot of time 'double-banking' in every aspect of the job.

The vast majority of staff took the new recruits under their wing, although there was some degree of resistance to the new staff who had no sea experience. However, once it was clear that the standard of operating was as good as (if not better) than some of the old hands, such resistance quickly diminished.

After initial training, there was the probationary year during which staff had to pass a 27 words per minute Morse test, a station 'walk-around' test, and a French examination. One recruit came from Mauritius, whose main language was French. He was somewhat dismayed to learn that he was not exempt from the French examination as he did not possess the required exemption certificates. In the latter years, the French examination was not given such importance and was eventually dropped.

One lovely story involved George Davison, an experienced and kindly R/O who enjoyed working with the new recruits. On SLT duty one day, he was showing one young recruit how to complete a 'reply-paid' voucher which enabled the recipient to reply to the incoming message. The young man was somewhat concerned that the voucher was actually larger than the SLT envelope. "Ah, yes, big problem," said George. "A tricky one. What to do? Ah yes, we fold it!"

Another episode (again involving an SLT) occurred when double-banking. Both the senior R/O and the trainee took down the same message, both on individual typewriters. Unsurprisingly the trainee's form contained a few layout errors, but the experienced R/O's version was perfect. The young recruit was advised to dispose of his version but unfortunately, he ripped up the wrong one – so the whole SLT had to be retyped.

Work on the new station building had started at the end of the 1970s, with new message handling procedures being introduced for both W/T and Radiotelex services. The W/T system removed the need to transcribe telegrams directly onto typewriters; 'Trend' teleprinter terminals were installed on each W/T point and linked to a central computer processor (Honeywell 606) which would automatically handle and route both to-ship and from-ship traffic.

W/T console in the new station showing the 'Trend' teleprinter in use at the time.

The 'Trend' terminals needed plenty of maintenance. They used inkjet printing technology, and the print heads needed regular cleaning and replacement. The keys themselves were quite fragile, and many were destroyed by fist-thumping of frustrated R/Os trying to work a ship whose Morse was not of required operating standard. Eventually, these were replaced by computer keyboards and screens which were much easier to read, maintain and operate.

The old building was in poor condition; rudimentary heating and ventilation systems were still in place, and time had taken its toll on the 30-year old consoles, with some switches becoming unreliable and inefficient.

'C' Wing, in particular, was showing the effects of cigarette smoke and was still very much the 'smoking wing' within the station. A layer of cigarette smoke was often to be found above the 16 W/T consoles throughout the day (and indeed sometimes on the night shift). The aerial amplifiers, in place since 1948 were showing signs of deterioration.

The Radiotelex section, just before the move to the new building.

The move to the new building took place in stages; Radio Telephony was the first service to move in, with gleaming new consoles, receivers, and much more space to move around. This was followed by the W/T service, initially on 22 MHz only, followed shortly after by the services on the other bands, and finally the radio telex service, once the new automatic system had been tested and deemed to be working satisfactorily. The interim stage of the radio telex system commenced in July 1984 and became fully automatic during 1985.

A purpose-built landline area, in a room next to the main operations area, was built, which consisted of 3 incoming telephone booths for transcription of dictated messages; a bank of Post Office type 15 telex machines; numerous 'Trend' telex machines with reperferator facilities in case of computer malfunction; and a private wire/services area. There was also a 'Wendy House' for the landline overseer, which was rarely used. The new building was to be a 'no smoking' zone, although those who wished to smoke could venture into the garden area if necessary.

A link to the Admiralty Communications Centre (COMMCEN) was maintained over the TARE network, where "TARE Checks" were made every 30 minutes to ensure a reliable connection. Any ZBO messages received at the station were passed to COMMCEN over this network although these were few and far between. However, some from-ship traffic (mainly SLTs and service messages) were received on this link.

The 'Burnham Message Handling System', originally trialled in the old building, became fully operational when all services were installed in the new building during 1983. An abbreviated description of the system and its functions is reproduced below, firstly in the to-ship direction;

"Most incoming telegrams will be accepted at Portishead Radio via telex, telephone or private wire. All these telegrams will be forwarded to the TRC (Telegram Retransmission Centre) in F31 (International Telegram Standard) format, in the case of Portishead via the to-ship concentrator. Radiotelegrams handed in overseas will route directly to the TRC via normal international circuits.

The TRC, on identifying 'PORTISHEADRADIO' in the last line of the address is programmed to compare the ship's name, which must be contained on the penultimate line of the address, against the entries in the SNF (Ship's Name File – a list of ships names and call signs, maintained by a dedicated SNF Officer at Portishead). If a match is obtained between the name of the ship on the radiotelegram and an entry in the SNF, the TRC adds the information contained against that file entry to the bottom of the telegram.

The TRC then automatically routes the telegram to Portishead Radio over one of the four lines available for this purpose.

At Portishead, all telegrams incoming from the TRC are presented to the processor (Honeywell 606 as mentioned above), which is programmed to check the validity of the information line inserted at the end of each telegram. If the validity of the Ship's information line is verified, the telegram is stored in the processor and the information contained in the line is utilised to process the telegram. On the other hand, if the Ship's information line does not comply with the requisite parameters the telegram is rejected to an Edit Position for manual handling.

The processor utilises the information contained in the Ship's information lines of the stored telegrams to construct separate traffic lists for WT, RT and RTT fitted ships. These lists can be produced, on-page copy and 5-unit tape, either automatically when a radio transmission is due, or on-demand.

All traffic lists will be printed out on the Traffic List Printer where, if required, facilities exist to produce a 5-unit tape. Additionally, the RT Traffic List is printed out at half-hourly intervals at the RT Consoles.

In the from-ship direction, the Message Handling System caters for seven Search Points and up to 44 work positions. This provision may be subdivided into 6 groups, each group having access to not more than two Search Points or twenty work positions.

Each Search Position will be provided with a 'Trend' printer which will have an interface with a processor. On hearing a calling ship, the Search Point officer will type in the following information:-

i) Ship's call sign

ii) 3 digits indicating the last 3 figures of working frequency (QSS). Optionally this may be followed by 'C' or 'D' to indicate the Portishead Frequency (GKC or GKD) being monitored by the ship.

iii) 3 digits indicating the optimum directional aerial. Optionally this may be followed by the letter 'A' to achieve a degree of priority."

The from-ship concentrator assembles the telegram into full F31 format and queues it for onward transmission via the appropriate national or international network. Companies with 'Private Wire' access, such as Niarchos, Bracknell Weather Centre, and the Admiralty, each have their own 'routeing codes' which by-pass normal delivery methods ensuring almost immediate delivery."

This 'basic' message handling system lasted, with minor enhancements and amendments, until the end of the service in 2000. When the 'Trend' machines were replaced by computer keyboards and screens at the end of the 1980s, the handling of radiotelegrams remained virtually the same, although the Honeywell 606 processor (housed in its own room) was replaced with a processor far less cumbersome.

A combination of new technology and inquisitive staff could have interesting results, as R/O Dave Drew recalls from the early days of the new station:

"In the new high-tech landline room there was a computer terminal into which commands could be typed to generate various reports. Its precise function or title escapes me now but I recollect that there were quite a number of different commands, those below ten each being prefixed with a zero.

On the night in question, I was sitting beside this computer terminal at about 0200, trying to keep awake and look busy, when I happened to look at the list of commands and noticed that there was no command zero-zero.

I swear that I tried to resist the temptation but eventually I felt compelled to type in zero-zero and press enter. Nothing happened. About half an hour had passed when the landline door suddenly burst open and in rushed the overseer. Whoever it was he was looking a bit baffled and concerned and wanted to know if I'd touched anything or noticed anything unusual in the landline room. Being an honest sort of chap I, of course, replied that I hadn't touched anything and the landline room was functioning normally (the last bit being true).

The overseer quickly departed the landline room muttering something about chaos in the main operating area so I followed him out.

In the central control area, every machine was chuntering away spewing out yards of paper. There were traffic lists by the score, miles of five-unit tape and I can still see another machine printing out copies of messages that we had sent and cleared several days before. After much scratching of heads and a good deal of foul language, the general consensus of opinion was that the 'bloody computer' had lost its marbles and that the on-call engineer should be sent for.

Work was still going on to restore the system when I went home at 0800. I found out afterwards that the engineers could find no fault on the system and that their best guess was that the computer had been given a command that it didn't understand and had gone through its entire repertoire in an effort to please. Thankfully, no-one could trace where the misunderstood command had originated from; it was just a computer glitch."

Use of the new message handling system no longer required log books to be maintained, as details of each contact would be electronically held. However, some staff preferred to maintain their own personal log books.

The outbreak of the Falklands conflict in 1982 brought extra work to an already busy station; overtime was laid on for those who needed it to cover this extra work (for which many of us were extremely grateful). Many 'STUFT' (Ships Taken Up From Trade) were regular users of the service, and all (with one notable exception!) were fitted with Radiotelex, thereby making communication easier for all concerned. The exception noted above was the vessel *'Balder London/GURK'* who had to send and receive all messages (which were encrypted) by Morse code. Bearing in mind many messages exceeded 1000 groups of five letters, the poor chap on board certainly earned his money. Royal Naval staff were seconded to the station to 'vet' messages to and from vessels involved in the conflict, and to ensure correct procedures were being followed.

The Falklands had an existing point-to-point radio link established from the receiving station at Somerton. Fearing the possibility of Argentine jamming and/or monitoring of radio communications between the Falklands and the United Kingdom, it was thought advisable to recalibrate the link between Stanley and the station at Rugby and switch to an alternative station at Portishead. The Governor had agreed to this move under the strict proviso that, 'they kept some link open with the UK'. That demand went unfulfilled, and the connection with Portishead was never established.

The very first night of the conflict was a trifle confusing; staff were advised to monitor the calling W/T frequencies for 'FTF' (Falklands Task Force) vessels before it was realised that all vessels (bar the one previously mentioned) were fitted with radio telex. Once this matter had been realised, the radio telex room became a 24-hour hive of activity, as 5-unit tapes received on the telex reperferators were passed from the landline room ready for onward transmission to the ships concerned.

Direct communication with ships was only possible if they were located North of Ascension Island – as soon as they passed the island, all (non-emergency) traffic was sent via the secure Naval networks, with HF radio silence being imposed. This was, of course, a frustrating time for the families of those serving on the vessels, although telegrams could still be forwarded via the Admiralty for onward delivery over their own network.

Events were closely monitored by the staff, who by this time had built up excellent working relationships with the radio personnel on both the Royal and Merchant Navy vessels.

At the end of the Falklands War, Portishead returned to 'normal' communications, although ships returning from the South Atlantic were still making numerous calls home once Ascension Island was passed. The BT Staff Magazine 'Telecom Today' published a short resume of the activities of Portishead, which read;

"Portishead Radio, BTI's long range Maritime Radio Station had a busy time as the ships came home from the conflict in the South Atlantic. The station provides HF radio telegram, telex and telephone circuits to ships at sea in all parts of the world.

The start of the conflict saw HF Radio Silence imposed on the fleet together with the procession of passenger liners, tankers and car ferries as they made their way south, and the only ships in the conflict zone able to use Portishead for both-way telephone calls were the hospital ships – the cruise liner *'Uganda'* and four Royal Navy survey ships, converted to their new role. However, there was no dearth of Morse and teleprinter broadcasts for the many ships still using these systems.

At the cessation of hostilities, homeward bound vessels were able to use HF Radio once past Ascension Island and the unnatural quiet of the previous months was shattered as ship after ship called in with telephone traffic. Every spare transmitter and console was pressed into service to deal with the flood of calls as soldiers, sailors and merchant seamen tried to cram three months of experience into a three-minute call. That the demand was met is evident from the signals, letters, ships' crests and photographs received from Task Force ships as they finally reached their home ports.

The *'QE2'* is now back at sea, shortly to be followed by the *'Canberra'* and *'Uganda'*, back in their designed role and pumping welcome holiday traffic through Portishead Radio once more. Sadly some of the ships, whose name and call signs had become so familiar to Radio Officers at the station, will never come north of Ascension again.

Task Force communicators, able to be released from their ships, have been invited to a social evening at the station to be held in September – this will enable both ship and shore side operators to put faces to the voices they have worked on the long haul north".

The Falklands Task Force Radio Officers with some Portishead Radio staff.

1982 also saw the introduction of the Ocean Weather Service (OWS), a system whereby dedicated weather reporting vessels (British, Dutch, Norwegian and Russian) each located at specific positions in the Atlantic Ocean, would send weather reports to the UK National Weather Centre at Bracknell. Bracknell themselves used to operate this service but found it more cost-effective to use the existing service at Portishead. This was achieved by use of the Radiotelex Service, with Portishead Radio having the ability to 'poll' the vessels concerned to request the reports at scheduled times, 24 hours a day. Dedicated radio telex frequencies, using the 'GKL' series of call signs, were introduced for this service.

Vessels regularly participating in the service included:

- *'Cumulus/GACA'* – position 52.00N 20.00W (Station L).
- *'Polar Front/LDWR'* – position 66.00N 02.00E (Station M).
- *'Ernst Krenkel/EREU'* – position 52.45N 35.30W (Station C).
- *'France II/FNEL'* – position 47.00N 17.00W (Station R).

The following USSR-flagged vessels rotated on Station C as required:

- *'Victor Bugaev/ERES'*
- *'Musson/EREA'*
- *'Passat/UZGH'*
- *'Georgiy Ushakov/ERET'*

Each participating vessel would provide Portishead with a schedule, with primary and secondary operating channels, on which the ship's equipment would be set. This tended to work very well, although the vessels provided to the scheme by the Soviet authorities tended not to follow normal procedures at times, meaning lapsed schedules and the occasional call from Bracknell requesting reports.

Vessels would only be 'on station' for a finite period, and weather reports were also taken from replacement vessels 'on passage' to and from the designated positions. Schedules were updated regularly to take into account radio propagation conditions and times of day.

Initially, the operating point at Portishead was in the old radio telex area at the end of 'A' wing and when the new building was completed, it was relocated in its own section of the radio telex operating area.

R/O Roger Marshall at the original OWS position in the old station, 1982.

BT regularly operated a stand at the larger 'Boat Shows' such as the ones at Earl's Court and Southampton in order to promote the various communications services on offer. Many commercial ships were asked to participate in demonstrations from the stand, as Tim Strickland remembers:

"At times of the Boat Show, it was usual for the R/O at Portishead to ask if anyone would be willing to be connected to the BT stand at the boat show and answer questions regarding your ship, its route, cargo and life on board, etc. from visitors and the BT staff at the show. I always volunteered to be connected to the boat show stand and remember once being asked to describe life on board at that moment. As it happened, we were in a period of beautiful weather off the coast of West Africa so I gave a long descriptor of calm seas, blue sky, unbroken sunshine, a swim in the pool and steak and chips for dinner. There was a gap when I finished speaking and then the Portishead R/O at the boat show stand came back with "and you are paid to do this work, is that correct?" The payback for taking part in the boat show was the reward of a free of charge three-minute call home which usually stretched to five minutes with no charge."

Christmas at Portishead was always busy for obvious reasons, but some recollections are worth mentioning. Neville Edwards recounts:

"Christmas at GKA in the 1980s could only be described as 'manic' - with R/T being the busiest service by far. It was all hands to the pump in the R/T Coordinator's office, taking bookings (and in many cases refusing bookings when a 'book in advance' system was in operation). Also handling numerous calls from inebriated spouses on Christmas Day and learning how to handle their woes. Sometimes it was like working for the Samaritans."

Portishead R/O Graham Powell agrees:

"I was on telephone duty one Christmas morning when an obviously distraught lady rang up demanding to speak to her husband on board ship. She even threatened to commit suicide if she couldn't be connected. At the time R/T calls on Christmas Day had to be pre-booked. I sought guidance from the duty overseer (who shall remain nameless) who unsympathetically advised me to tell her to carry on – we're far too busy."

Certainly in the early 1980s ships were only allowed 30 minutes per turn - at the end of the 30-minute spot they had to go to the back of the queue, although some high traffic vessels did manage to arrange dedicated circuits.

The 0800-1300 watch was by far the most popular - volunteers were always sought for the 1300-1800 and 1800-2300, although those in financial need were available for anything going. An 0800-2300 shift always helped to pay the bills, although the more hardy R/O would try for a 2300-0800 and back on at 1300-2300 shift to get the maximum allowance.

In the old station, the 5-unit telex tape would be hung across the roof of the RTT/Landline section to give the semblance of decoration - the control room would have a model Santa Claus strung up by his neck, and a Christmas tree mounted on top of the traffic 'carousel'.

Of course, the welfare club was always popular, and this was always well decorated. A Christmas party was always arranged for the children of staff, although one child was heard to question why Santa's breath smelt of beer and wore trainers.

Bob Singleton recalls:

"One Christmas Eve, I think, which fell on a weekend, Mike McCarthy was night overseer and the Welfare Club had had the kids' Christmas party earlier. Someone had put in an appearance as Santa and left his suit in the station. What better way for Mike to wake up the first lot of bods snoozing on their break in the rest room that Christmas morning than appear as Santa with a merry Ho Ho Ho!"

In the new station, especially when the other main BT department moved in, it was definitely a building of two halves - enter the non-radio station area and you would be blinded by flashing lights, tinsel, Christmas trees and presents everywhere. Open the door to the GKA section and it was like entering cold-war era East Berlin - no sign of festivity and the atmosphere of 'humbug' prevailing. This was of course when the 'writing was on the wall' and it was clear the station was in its final throes"

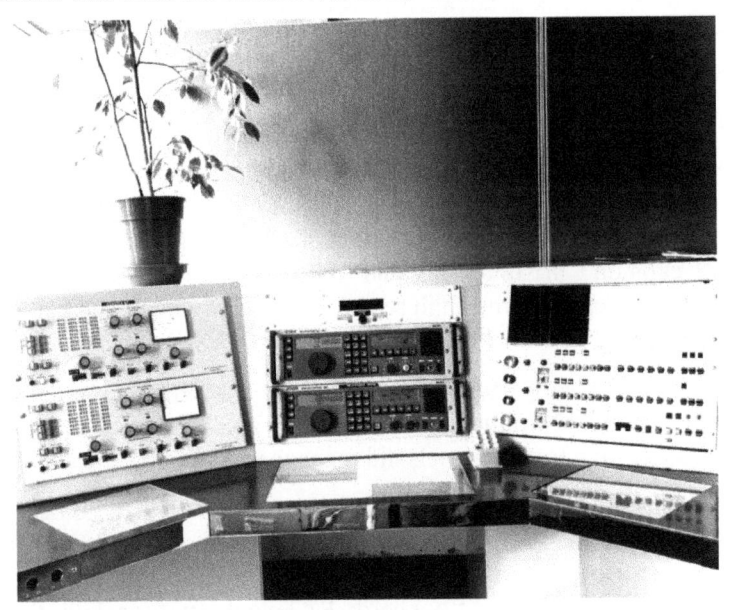

A gleaming new R/T console, ready for operation.

The move into the new station was completed in 1982, and use of the receiving aerial site at Somerton commenced. The aerials and receiver control units were linked to the radio receivers at Somerton via a microwave link via Pen Hill, on the Mendips, and the redundant aerial system at Highbridge was demolished in early 1983.

One of the Highbridge aerial masts after removal in 1983.

The aerial system at Somerton was clearly defined by the description provided in the official station 'Equipment Notes' at the time;

"The HF aerial system of horizontal rhombic aerials accepts signals in the frequency range 4-30 MHz and provides maximum directivity within a 360 degrees coverage.

The rhombic aerial has a suitable direction response throughout the required frequency range and a relatively uniform impedance-frequency characteristic which is easily fed into the receiver distribution system.

24 aerials provide for reception above 10 MHz (HF) and 12 aerials provide for reception below 10 MHz (LF). On the HF aerials, average bandwidths of 18 degrees and on the LF aerial average bandwidths of 36 degrees are secured. The 24-position switch on each console provides, on alternate positions, either a single HF rhombic or a combination of HF and an LF rhombic. The frequencies are obtained automatically by the use of high pass and low pass banded filters.

It may be noted that the layout of aerial direction is not arranged to provide equal azimuthal coverage throughout the 360 degrees. This is because Somerton was originally designed for point-to-point working with fixed land stations. When it was taken into use for maritime purposes the aerial ring was altered slightly and additions were made in such a manner that adequate 360 degrees coverage could be secured.

The omni-directional aerials are a combination of 4-6 MHz triple dipoles and 8-12-16-22 MHz stacked quadrant dipoles."

Provision was made for vessels in the Pacific, similar to the 'Pacific Watch' which operated in the old station. This was known as the 'Sector Watch' which monitored Channel 2 in the 4 to 16 MHz bands and Channel 5 in the 22 MHz band. Use of the sector watch was on instruction from the Radio Overseer when radio propagation conditions were favourable on both long path (LP) and short path (SP) routes, but in reality, it was seldom used. The sectors were defined as below:

Sector 1 – aerial bearing between 216 and 252 degrees

Sector 2 – aerial bearing between 252 and 288 degrees

Sector 3 – aerial bearing between 288 and 324 degrees

Sector 4 – aerial bearing between 324 and 000 degrees

Sector 5 – aerial bearing between 000 and 036 degrees

Sector 6 – aerial bearing between 036 and 072 degrees

Sector 7 – aerial bearing between 072 and 108 degrees'

The BTI publication 'Long Range Maritime Radio Services' explains the service clearly;

"Channels 2 and 5 in the bands 4-16 MHz and 22 MHz respectively are assigned for use by ships working Portishead Radio from long distance. Unfortunately, the channel is also shared by other administrations for general purposes.

Watchkeeping, channel by channel, will be on a sequential basis by the GKB Radio Officer, as with other channels.

It is to be expected that short path routes will apply on the highest band in use, and to some extent on the next highest band. On the lower band, it is probable that most signals will be on long path routes. On the short path routes, the directional receiving aerials become critical and signals may not be heard on omni-directional aerials.

When listening on routes that require critical selection of directional aerials the Radio Officer may add a Sector Watch indicator to his tuning signal emission and this will indicate the sector under search e.g. de GKB S6 in the 16 MHz band will indicate Sector Watch on channel 2 in sector 6. If you are in another area and decide to chance a call you may not be heard.

If you hear the appropriate channel indicator without the sector indicator then you may assume that the Radio Officer is covering all sectors.

This service will provide an almost continuous watch covering from the East coast of South America and the West Indies clockwise to the central Indian Ocean. Ships may use other channels for calls from these areas. Please note that our receivers will be covering up to 19 frequencies in one channel on 22 MHz – it is not a single frequency operation".

It was obvious that although the station was trying its best to provide as good a coverage as possible from the area of concern, a combination of radio conditions, equipment on board and the radio receivers at Portishead made this extremely difficult.

One R/O at sea complained about the service, however;

"On my present vessel trading between Alaska and Japan I find contacts with Portishead somewhat of a struggle, especially when during an opening of a few hours between 1900-2100 GMT when GKB6 is QSA5 – but has his QRL tape running for approximately 65 minutes of the time. Take into account the time at the start of the opening that his signals are improving and at the end when they are fading, plus the fact that JCS is sending his traffic list (albeit half a kHz down), during the opening, no wonder I get anti against the system. So where is the Sector Watch or is all the published information all useless and does it require a deletions correction?"

Don Mulholland, Station Manager, replied;

"The R/O is not really complaining about an apparent lack of coverage on the Sector Watch channels because it is available on all bands open on a continuous basis. When the QRL tape goes on during a search period this implies that the total delay on all bands is in excess of 30 minutes and rising. There is no alternative but to close search and go QRL whilst the ships on turn are worked off.

Table 2 : Sector Watch

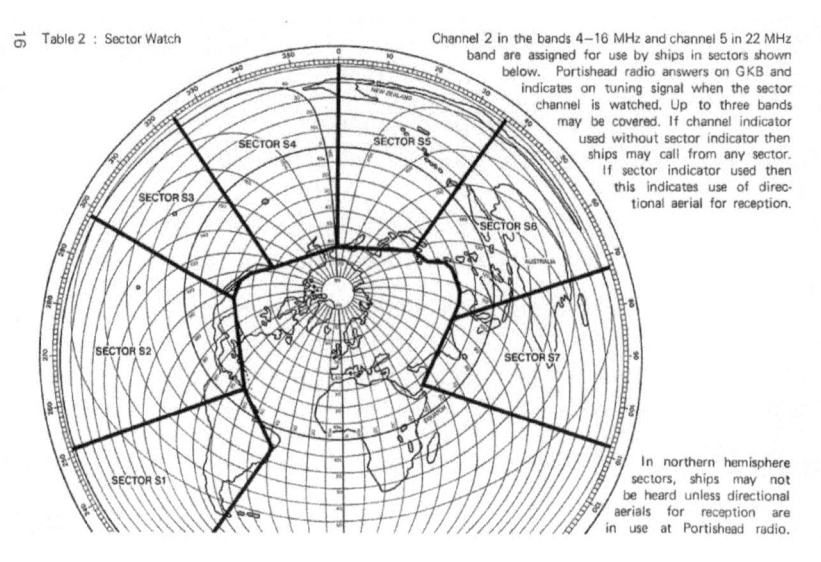

Channel 2 in the bands 4–16 MHz and channel 5 in 22 MHz band are assigned for use by ships in sectors shown below. Portishead radio answers on GKB and indicates on tuning signal when the sector channel is watched. Up to three bands may be covered. If channel indicator used without sector indicator then ships may call from any sector. If sector indicator used then this indicates use of directional aerial for reception.

In northern hemisphere sectors, ships may not be heard unless directional aerials for reception are in use at Portishead radio.

The 'Sector Watch' as explained in the "Guidance for Radio Operators' publication.

We are concerned about the difficulties facing Radio Officers at sea particularly during this critical evening period but because of economies placed up on us there is indeed little scope for improvement. However, efforts will continue to be made to reduce delays and thereby keep the search watch going.

Leaving that aside, the R/O operates in a very difficult part of the world for communications with the UK/Continent. Even at this period of the sunspot cycle, a ship off Alaska is lucky to get through at all".

Obviously, the service was reliant on reliable radio propagation conditions, and as the W/T service declined at the end of the 1980s, the Sector Watch was lapsed.

It is interesting to look back on some correspondence from some seagoing Radio Officers published in the REOU 'Signal' magazine at this time. It does seem that there was some discontent with the W/T service offered by Portishead during the night shift.

Before dedicated Search Points became an H24 operation in the new building, it was normal practice to perform a 'search-and-work' operation, whereby individual R/Os would listen for calling ships and work them immediately rather than line them up for working. Occasionally, when many ships were calling at the same time, a small list of ships would be compiled by one individual who would work them sequentially. This method of working would lead to many cases of 'DE GKB QRL' being transmitted by GKB, advising ships that the station was busy at this time.

Such comments included;

"The problem appears to be by the station being overworked and undermanned. This means that, apart from a skeleton service provided at night for a few lucky vessels, the station is only available from 0800 to 2200 LT. This is a ridiculous situation to be in, when a service is in great demand, but due to lack of staff, an adequate servicer cannot be provided, despite large numbers of R/Os being unemployed. The station is overworked due to the number of foreign ships using GKA as a 'convenience' station i.e. reliable and cheap. The station is undermanned because BT cannot afford to pay for more R/Os to be employed.

"May I say first, that I consider the service provided by GKA first class, during the hours 0800-2200 approximately. However, GKA just seems to die, between the hours 2200-0800. I appreciate that there may be staff shortages and economies etc. but, with just a little imagination, the service during those night hours could be well improved with regard to the frustrating hours spent being bombarded with the QRL tape. I have experienced tapes running for well over 2 hours with the annoying situation that when GKB eventually listens he more often than not works the first ship heard then switches the QRL tape back on, again for very long periods. When a listening break occurs surely a QRY list could be built up over a listening period of say 5 minutes or more, as I'm sure ships would rather be QRY20 (or more) with GKD/C than again listen for the QRL tape break which could be hours away, and then GKB working just one ship once more?"

One more letter was even more scathing;

"What goes on at Portishead from midnight to dawn bears little resemblance to anything in published articles; QRL on all bands most of the night – operators who think they were appointed by the Almighty to deliver sermons on their version of the Regulations, the standard of everybody else's Morse, the construction of your telegram – missed their true vocation, should be in the police force. One sits all night listening to a QSA5 plus QRL on 12 MHz, calls in desperation on 6 MHz, and receives a lecture about 'selecting the optimum frequency'!!

The congestion at Portishead during the day is directly related to the absence of any useful night service. If Portishead is to be an 'office hours only' station, better lose it down altogether at night and save fuel. Give the traffic to the European stations like Norddeich/DAN, which are willing to work at night".

The Station Manager at the time, Don Mulholland, was compelled to reply;

"The point is taken that no-one likes to sit listening to a QRL tape especially when it may seem that no ships are being worked. Let us examine the situation.

I have looked at the from-ship radiotelegram traffic over 21 days and produced daily average totals for the periods 0000-0800, 0800-1600 and 1600-2400. Against this is shown the daily staff hours used.

It may be noted that:

(a) In the period 0000-0800, some 200 items were received and this did not include transactions such as to-ship radiotelegrams, SLTs, TRs, and services.

(b) The night productivity was highest at 3.2 items per hour.

So, we may be satisfied that the night staff do not hang about and that far from being idle, they actually work ships".

In the event, the problem disappeared with the traffic handling methods in the new station reverting to 24-hour Search Point operation.

The new station was certainly needed, with the ageing equipment and buildings of the old station requiring a high degree of maintenance. The operating wings were demolished (with the exception of D wing, which became an office area), and a new corridor to the staff restaurant and welfare block was built. In addition, a link to the new station (with a keypad-operated security door) was erected. Staff still had to enter the building via the original front door and use the new corridor to access the new station.

In addition to offices for the Station Manager, Assistant Station Manager, Chief Engineer, Administration, Writing Duty and Accounts, there was a large W/T operating area of some 48 consoles, an R/T section of 12 consoles, a dedicated landline room, and a radio telex area, which now offered an automatic service with minimal operator assistance. Within the W/T area was the overseer's area, adjacent to the broadcast section where traffic lists and other broadcasts were compiled and scheduled for automatic transmission.

Naturally, a large VF room and Engineer's Workshop were part of the new station area, as well as a comfortable rest room for staff.

The old and the new. The new station building viewed from the station entrance.

There was also a small garden area (designed by one of the staff members, Mike Howley) where staff could relax during breaks and rest periods.

Each console was provided with a Racal MA1075 receiver front panel (the actual receivers were at Somerton and remotely controlled from Burnham), a transmitter selection unit, a 'Trend' teleprinter, an intercom system and a choice of Morse keys; a custom-made key manufactured by the Rugby Workshops, based on the Marconi PS213A, and a Katsumi EK-150 electronic key. The Katsumi key was chosen after trials of other electronic keys by the staff, which included a model made by Heathkit and a Samson ETM-3C.

The R/T consoles used the same receiver and aerial systems, but had a comprehensive collection of controls on hand in order to provide the highest quality link available. Although looking complicated, the actual operation was relatively simple.

The consoles were designed to allow two circuits to be controlled from each position. The circuits are designated system one and system two. Each system consists of a remotely controlled transmitter, a remotely controlled receiver, and a terminal unit. The operator controls the systems with an operator's control panel, with controls for filters, amplifiers, and other circuitry to enhance the landline connection. A direct link to the International Telephone Exchange at Glasgow was also installed to assist with bookings made from outside of the United Kingdom. All calls were manually timed and charged by the minute, with a 3-minute minimum duration, although it was entirely at the discretion of the Portishead R/O what to charge, taking into account interference and call quality.

An evocative view of the new station shortly after opening in 1982.

The Ocean Weather Service was the last part of the station to leave the old building, and at the end of 1986, the whole OWS system was incorporated into the new BART (Burnham Automatic Radiotelex) system, thereby dispensing with the need for manual interface. All vessels were automatically polled at their scheduled times by the BART system, which involved setting up schedules manually by a radio telex operator. Once this was done, the system polled the vessels at these times and on the scheduled channel(s) until manually amended.

Looking at the traffic figures for 1981/1982 it is clear that the station was still a world leader in handling maritime communication. For radio-telegrams, 22,282,050 words were handled at Portishead, compared to 24,798,741 the previous year, a decrease of 10.15%. HF radiotelephony minutes were slightly down on the previous year (986,617 from 995,317), but HF radio telex minutes increased from 509,736 to 545,086.

A major culture change occurred when the Post Office Maritime Radio Services became part of British Telecom PLC as part of the privatisation process in 1981; The radio services were assigned to BTI – British Telecom International – and as a result became accountable to BT's shareholders rather than to the company itself. This brought the service under the accountant's spotlight as the service, whilst being vital to ship communications and for safety of life at sea, had never been regularly profit-making.

Staff were, however, pleased to receive some free shares in the new company, and given the opportunity to purchase further shares at a substantially discounted rate. Many took advantage of this offer and made an excellent profit when selling them some years later.

However, this change did not affect the day-to-day working of the station, which continued to provide the usual excellent service.

The new 'broadcast' position with R/O Ian Benfield feeding 5-unit tape through the reader. Receiver and transmitter selection unit seen above.

R/O Des Thompson at one of the R/T Consoles.

View of the R/T consoles in the new building.

A spectacular view of the Microwave tower at the Highbridge site at sunset.

The popular Ship Letter Telegram (SLT) service was withdrawn at the end of 1982 following the introduction of the new computerised message handling technology, and despite many protests, it was never re-instated. The REOU felt strongly against this even advising members to continue to receive to-ship traffic from Portishead but from-ship messages to be sent via the most convenient coast station. This would have the effect of not swelling GKA's coffers and also to see how effective it would be in the case of a future confrontation with the Owners. It seems there was initial success with a large amount of traffic being diverted, in one case over £1,000 worth of revenue within the first four weeks.

BTI responded by introducing a Radio Telex Letter (RTL) service, but this was only available to vessels equipped with radio telex or satellite telex equipment. The REOU pointed out that this would not be available to 80% of British-registered vessels and was, therefore no viable substitute.

A further enhancement came with the Greetings Telex Letter (GTL) service where an RTL was inserted into a requested greetings card before being posted to the destination. This did, of course, incur a further charge. The design of the requested card could be chosen from a list held in the radio telex databank.

The automated message handling system dispensed with the telephone delivery of telegrams received after office hours. To counteract this, new value-services were brought to the station such as 'Telecom Gold' – a service where from-ship messages received after office hours would be diverted Telecom Gold's central mailbox to a customer's dedicated mailbox. The customer would then be alerted via a radio-pager to check his mailbox. This was very much a precursor to the 'Voicebank' service brought in a few years later to handle such traffic.

An internal memo dated 21 June 1983 and only recently discovered, shows how close the service came to closing in the early 1980s;

"It is becoming increasingly uneconomic to provide telecommunication services to ships by way of coast stations. BT would prefer to see all ships equipped with the equipment necessary to communicate through satellite links provided by Inmarsat and Marisat which correspond automatically with BT's large earth stations such as Goonhilly. This more modern service is not attractive to all shipping because:-

1. Tariffs are higher than for conventional coast station communication;

2. The ship needs expensive equipment to communicate with the satellites; and meanwhile,

3. A conventional coast station service is widely available in other countries which do not have an earth station like Goonhilly.

BT wishes to make its coast station service less attractive by increasing tariffs, with a view to phasing out the service in due course. However the owners of British-registered ships constitute a powerful lobby which opposes tariff increases and we are therefore considering a plan to raise only the gold franc tariff, a move which will discourage traffic with foreign-registered ships without prejudicing British-registered ones".

Plans were also made to remote-control certain coast radio stations as part of the Distributed Organisational Control (DOC) system, with Burnham being planned to be one of the main control centres, as explained in an article from 1982;

"Land's End Radio, the maritime communications station that played a significant part in the Penlee lifeboat disaster rescue bid, likely to end operations in its present form within three years. British Telecom said yesterday it is one of 11 such stations around Britain under threat as a result of a Whitehall reorganisation plan. Under the new plan, all shipping communications from Britain will be concentrated at two stations Burnham on Sea in Somerset and Stonehaven. The 21 staff at Land's End were told yesterday their station will be remotely controlled from Burnham. The formal notification to the staff left them in no doubt that their number will be decimated, and that the future of the maritime station at Ilfracombe also in doubt. Mr. Clive Knott, a radio section committee member in the Union of Communications Workers, said: "It seems likely that of the total staff of 151 radio operators there will be a cutback of 90. The Land's End station has been in operation round-the-clock since 1913".

In reality, a version of 'DOC' did take place later in the decade, with Land's End and Stonehaven becoming the control stations, the plans for Burnham being dropped. BT confirmed this in their internal communication of October 1986 which stated:

"The UK will be divided into two separate regions of four coast stations linked together by speech and data control circuits with medium range stations closing in January. This system of Distributed Operational Control (DOC) will mean that a calling ship will be linked to whichever station in its own region can handle the transmission soonest instead of queuing for a particular station which might be busy at the time. This means that for commercial purposes no one coast station will need to be manned on 24 hour a day seven day week basis, although BTI will still maintain permanent cover for the distress watch and other navigational services for the Department of Transport. The Department has decided that because of a decrease in the volume of shipping – and a desire to cut costs – it requires fewer stations to maintain the distress watch. This will be provided by a continuously-manned watch at Stonehaven in the north and Land's End in the south and using links to the nominated distress stations".

Meanwhile, changes were afoot with regard to telegraphy procedure at Portishead Radio; for example, the W/T traffic lists which were previously broadcast each even hour were changed to EVERY hour on the hour from 1st June 1983. A list of British call signs was transmitted first, followed by an alphabetical list of foreign call signs. This change was due to a study of calling rates from ships following traffic lists, which showed a rapid increase followed by a delay in obtaining working frequencies for Portishead to deliver the traffic. It was therefore hoped that this provision of these extra traffic lists would smooth the calling rate and permit a smoother flow of traffic with reduced delays. In point of fact, this new system worked extremely well and certainly assisted in the expeditious delivery of to-ship traffic.

On July 4th, 1984, the radio telex system at Portishead became fully automatic, enabling vessels to dial directly to any shore telex customer via the station without operator assistance. This enabled the station to compete successfully against foreign radio stations, particularly in Belgium, France, the Netherlands, and Denmark, which had also developed automatic radio telex systems. It was planned (by June 1985) to provide automatic store-and-forward of telex messages to ships, automatic through dialling and a reduction of the minimum chargeable time for each call from three minutes to one.

Traffic figures for the year ending March 1984 are staggering; over 15 million words of paid traffic were handled on W/T, over 725,000 minutes of radiotelephone traffic and almost 600,000 minutes of radio telex traffic were connected. About 180 R/Os were employed at that time.

The Ongar transmitting site was shut down in 1985, leaving only Rugby and Leafield as the only sites used for the maritime radio service. However, the Leafield site was closed a year later, meaning that all future maritime long range radio transmissions came from the Rugby site only.

Most of Rugby's Point-to-Point rhombic aerials were replaced with omni-directional aerials. Following this, the 30 kW transmitters were replaced with lower power 8-10 kW versions. Some of these transmitters were recovered from the Ongar site. A number of these were brand new Marconi H1141 fast-tune transmitters, which greatly enhanced the service as they could be tuned to any frequency within the 1.6 MHz to 30 MHz range in just a few seconds. Details of the aerials used from the Rugby 'B' site were as follows:

The new automatic Radiotelex system with R/O Ron Burt monitoring traffic.

- Spiracone Aerials (x 9): Omni 4-30 MHz
- Wide Band Cone (x 2): Omni 3-30 MHz
- Rotatable Log Periodics (x 3): 4-30 MHz
- Fixed Frequency Omni (x 8)
- Rhombic Aerials (x 7)

The Rugby 'A' site closed around 1992.

British Telecom International (BTI) produced an updated version of the "Handbook for Radio Operators' which had been previously published by the Post Office. This publication was the Radio Officers' 'bible' and was the ultimate reference work for regulatory and procedural information. Of course, BT used the book to promote their own maritime services, but the procedures were internationally agreed and would be used irrespective of what coast radio stations were being used.

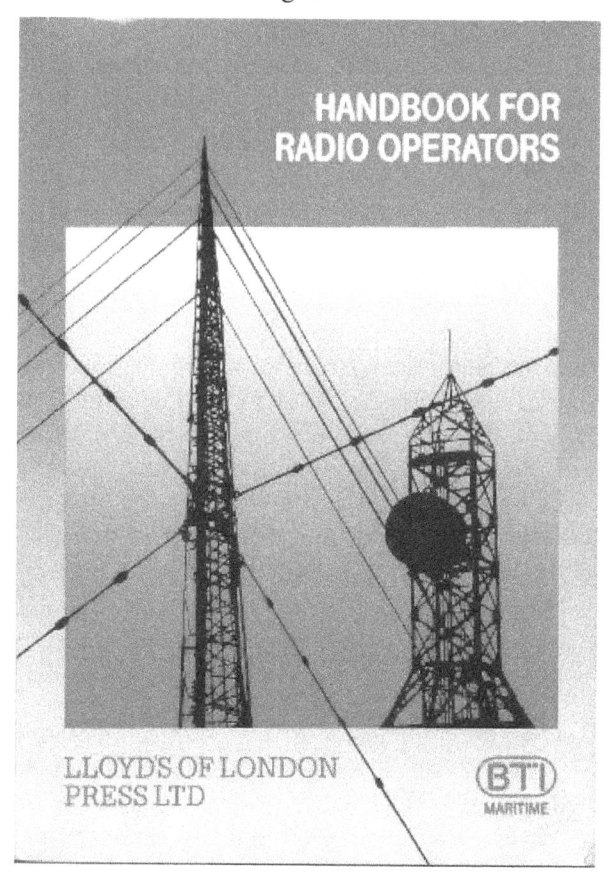

The 1985 'Handbook' produced by BTI.

As more vessels became fitted with modern radio equipment, BT decided to introduce new services to use Portishead Radio's infrastructure to handle traffic sent and received using Radiotelex equipment. In 1986, the 'Phonetex' and 'Voicebank' services were introduced. BT's publicity of the time stated;

"British Telecom International is launching two new round-the-clock telex services for ships fitted with Radiotelex equipment. About 18,000 vessels will be able to use the world-wide services. By using the new Phonetex service callers will be able to communicate by telex with ships at sea without needing direct access to their own telex machine. Customers simply phone BTI's Portishead radio station near Bristol and dictate their message to an operator, who will then telex it to the relevant vessel. Portishead is also extending its Radiotelex service to relay telex messages from ships to telephone answering machines in this country. The station will accept telex messages from vessels and then telephone the contents to the recipient. Messages can be sent to the customer's bank— BT's message manager service — or to an answering machine. For the Phonetex service, customers can dictate messages by phoning Portishead Radio on (0278) 781111. To use the Voicebank service ships should telex their messages to Portishead using the code VBTLX, stating the customer's telex number and Voicebank number".

BT was keen to promote the increase in satellite traffic (whilst still extolling the virtues of the terrestrial service) as an extract from an article in 1986 confirms:

"In the last two years, telephone traffic passing through Goonhilly coast earth station and the terrestrial coast radio stations has increased by 56 per cent and telex traffic by 86 per cent. There is still a significant use of the more conventional HF, MF and VHF radio services, and to meet the requirements of this sector, BT's programme of innovation has led to the automation of its long range and medium range radio telex service, trials with a medium-speed data service on MF and VHF and implementation of a coastal radio station network rationalisation known as Distributed Operational Control."

Such automation as mentioned above commenced in 1986 as BT announced:

"Telex messages for shipping on the high seas can now be received and stored, and relayed later - all automatically - through a new computer-based system installed at British Telecom's Portishead long range radio station in Somerset. Previously, storage and later transmission of telex messages was done manually by operators. Now, once a vessel is ready to receive the telex message, the new system will automatically transmit. All ships need do is enter their own watchkeeping arrangements in advance to the Portishead computer's database for automatic transmission during pre-determined times. In this way, telex messages can be received on board ship within minutes. A land-based customer wanting to send a telex to a ship sends the message on a telex machine in the normal way. The message is relayed to Portishead, where it is held in the radio station's computerised store, and forwarded to the ship. Ships not supplying watchkeeping instructions are called on a regular basis by Portishead until messages have been successfully delivered. Users of the service can also now send multiple messages during a single call. This feature will benefit companies using modern telex terminals with memory and pre-recorded address list facilities. One exclusive feature of BTI's radio telex service is the Frequency Watch facility. This enables watchkeeping instructions to be sent automatically to Portishead. Up to ten instructions a day can be stored for a maximum of 21 days. BTI's charge for its long range automatic radio telex service is £1.60 a minute, excluding VAT. With the introduction of new facilities, calls are now charged in steps of six seconds."

This service was later expanded to handle delivery of messages to and from vessels fitted with Inmarsat equipment.

The 'Voicebank' service was only used by a few customers, but remained in use for some time. Again, BT publicised the service with a brochure, stating;

"The latest addition to the Cellnet range of cellular phone products is Voicebank, a message service which can receive and send voice messages. When cellular phone users are away from their vehicles, callers can be redirected to Voicebank to leave a message. After returning to their vehicles they ring their Voicebank 'mailbox', which plays back any message. Each mailbox will store 48 one-minute messages for 72 hours, and users requiring a larger message capacity can have an overflow into further mailboxes. Up to 10 callers can leave a message at the same time on Voicebank, so an engaged tone is unlikely to be received. The Voicebank service can be linked to British Telecom's radiopaging system so that when a message is left the recipient's pager 'bleeps' immediately to alert him. A further feature is the 'broadcast' facility, which allows a manager to transfer the same message simultaneously to each member of his staff or to selected members. The Voicebank mailbox can be activated from office or public telephones as well as cellular phones, so staff not equipped with mobile phones can still access their messages while out of the office. The user simply holds a pocket-sized remote control keypad to the telephone and presses his access code to receive his messages".

One-day radio telex familiarisation courses commenced at the station in 1988. Designed to suit the needs of both the leisure market and the deep-sea fleet, the courses could also be tailored for fixed stations using the Gateway service. The content of the course provided the attendees with a good working knowledge of the radio telex system as well as understanding the procedures in place at the station.

As part of a staffing review held in 1988, many of the R/Os below a specified seniority level were defined as being 'redeployees' and were recommended to seek alternative employment if at all possible. A few staff did indeed find work with other departments within BT but many continued to work normal duties in the station. It eventually became clear that the rate of change of decline in the service was not as quick as originally predicted, and the redeployee status was removed. A similar exercise took place in 1994 when it became clear that closure of the station was unavoidable, and many staff did indeed move to other departments.

The station manager at that time was philosophical about the future of the station however, as press reports relating to the future of Morse code stated;

"I would say that over the next decade it will sink into oblivion," said Ernie Croskell, who runs Portishead Radio near Bristol. The station, which employs 130 radio operators and is operated by British Telecom International, currently handles 1,000 messages a day in Morse for ships. In 1980, the traffic was more than 2,500 a day.

Mr. Croskell said demand at Portishead for Morse traffic is falling by 10 percent a year, with a corresponding increase in the use of the faster radio telex service. He said Morse is still favoured by Third World ships because it is cheap.

The well-manicured entrance to the station during the 1980s.

The manager said the decline of Morse is being accelerated by implementation of the Global Maritime Distress and Safety System, which will require the world's 70,000 merchant ships to be equipped with new technology by 1999. He continued by saying Morse distress calls will then be superseded by voice transmissions, with Morse keys being replaced by satellite equipment and high frequency radios."

Sadly, his predictions were destined to be correct.

A section of BT Engineers (responsible for many of BT's radio stations in the UK) took residence at the station building, along with the administration of BT's Ship Inspection Office, who were responsible for checking compliance with radio regulations on board ships.

1989 saw the passing of Portishead Radio's most famous employee – Sailor the cat, at the grand old age of 17. He was replaced almost immediately with another ginger tom, this time by the name of 'Sparks'. He used to spend much of his formative years in the drawer next to the telephonist's desk, and again became a favourite subject for the local and in-house press. Again, BT took the cat on their payroll (á la Sailor) and a couple of R/Os undertook the feeding duties on a rota basis.

Sparks – the last station cat.

As the 1990s approached, the rapid uptake of satellite equipment on board ships together with the advancement of communications technology, ensured that the future seemed bleak and uncertain for this famous station.

CHAPTER SEVEN

THE FINAL DECADE

Aerial view of the Highbridge site, early 1990s.

As the new (and as it proved, sadly final) decade began, BT decided to work alongside the rapidly growing Inmarsat services in order to attempt to prolong the life of the station. Expansion of the Phonetex service for delivery of messages by Inmarsat rather than radio telex was established, and the use of Inmarsat-C for communication with the competing yachts in the various high-profile races was increased.

Other technology was explored, such as the radiotelephone 'Autolink' service, which although not used at Portishead was administered from an office at the station. This was a service which allowed connection of radiotelephone calls on VHF and MF without the need for operator assistance. This service was launched in 1990 and used existing radio communication equipment with the addition of a low-cost modem. Payment for calls could take place through a UK telephone number or a BT Chargecard, as well as via a vessel's accounting authority. HF Autolink was trialled from Niton Radio in 1994, initially on 8 MHz and then on 4, 12 and 16 MHz using an HF scanning system. However, the system was rarely used and the service was discontinued.

The regular 'Guidance for Ships' Radio Officers' booklet was re-designed and replaced with a loose-leaf binder entitled 'Maritime Radio Services' during the 1980s, with updated pages being replaced on a regular basis. The original blue binder was produced during the 1980s, and the 'white' version of the binder came in 1990, being the final iteration of this publication.

The publication included details of procedures, frequencies, charges and dialling codes for both international telephone and telex, and was a very popular and useful guide to Portishead's services.

A further plan was published in 1991 which explained the strategy for the remainder of the decade. This included the installation of a DOC southern node at Burnham, a gradual reduction in the number of Radio Officers in line with traffic, installation of a DOC northern mode at Burnham when northern region traffic reduced to a certain level, and centralisation of all staff to Burnham when the distress watch and broadcast commitment ceased and commercial traffic reduced further. In the event no centralisation took place, and Burnham remained outside of the DOC network apart from the RTT service, which was centralised from the Burnham Radiotelex section.

Maritime Radio Services binders, late 1980s and early 1990s.

As part of the Radiotelex Coordinator's duties at Portishead, control of the UK's NAVTEX service was introduced. A report from the time clearly describes the role;

"NAVTEX transmissions for UK Coastal waters are made by three of BT International's Coast Radio Stations, Niton on the Isle of Wight, Cullercoats on the Tyne and Portpatrick, near Stranraer. The messages for transmission originate from the Hydrographic Office in Taunton, or from the Met Office at Bracknell and are sent by them on telex to the RTT

Coordinator at Portishead Radio. They arrive on one of his system management VDU screens where they are edited to the correct format suitable for transmission. This will include a time code. At the appropriate broadcast times, the messages are automatically sent by private wire to the transmitters for transmission. A NAVTEX receiver at Portishead enables the Coordinator to monitor broadcasts off-air. He also has available a general coverage receiver that can monitor actual transmissions. The transmitters at the Coast Stations are on standby waiting for the arrival of the data to be transmitted, and are brought up on line at that time".

A dedicated customer services office was set up at Portishead, tasked with promoting and seeking new services as well as assisting with marketing products and dealing with customer enquiries. Such an office had been in existence for a few years, but the role was enhanced and expanded to allow more flexibility in promoting BT's maritime radio services. This office was also responsible for setting up accounts for users of the aeronautical and Gateway services, as well as liaising with shipping companies to build mutually-beneficial relationships.

The NAVTEX position at Portishead. Note the thermal printer for broadcast monitoring.

BT was quick to promote the unique nature of the station, and used some examples of the station's business as featured in an early 1990s brochure;

Cold Calling: Radio Officer Giorgio Matranga showed real dedication to duty by spending a night in a deep freeze! It was part of his acclimatisation for a trek to the Arctic as Radio Officer for explorer David Hempleman-Adams.

Brrr – It is Ranulph: Another explorer of cold places, Sir Ranulph Fiennes, used Portishead to get his team's messages home from the most northerly habitation in the world. When his team returned from the remote Ward Hunt Island, they called Portishead "the best, friendliest and most efficient service in the world".

Branson's Pickle: Portishead handled the communications for businessman and daredevil Richard Branson's three transatlantic adventures – two in a speedboat and one in a hot air balloon. The station helped him keep in touch, through his triumphs and through his mishaps – like his balloon landing in the sea!

Crashing Out: Someone else whose ultimately successful journey was fraught with problems was microlight pilot Brian Milton, who made several unplanned landings while on an epic flight from London to Sydney. During the 12,000 mile flight, Mr. Milton kept in touch with his family and operation base using Portishead's HF aeronautical radio service.

That Sinking Feeling: When an American ship took on water and began sinking off the Azores, Portishead's staff alerted its owners, who were then able to get pumping equipment to the scene to save it from sinking!

It is interesting to relate the cost of radiotelegrams and radiotelephone/radio telex calls at this time; the April 1990 price list shows:

Radiotelegrams to ship: 53p per word plus £2.00 per message

Radiotelegrams from ship for UK delivery: 59p per word (minimum 7 words)

Radiotelephone calls to ship: £6.48 for first 3 minutes, £2.16 for each additional minute. A reduced rate was available between 2400 GMT on Friday to 2359 GMT on Sunday at £5.19 for first 3 minutes, £1.73 for each additional minute.

Radiotelephone calls from ship to the UK: £6.48 for first 3 minutes, £2.16 for each additional minute. No reduced rate available in the from-ship direction.

Radiotelex messages to and from ship: £2.05 per minute.

All prices ex-VAT at 15%.

There were, of course, extra charges applicable for calls made for destinations outside of the UK, and for radiotelegrams routed via foreign coast radio stations.

During 1992, another attempt was made to transfer the radio services from Burnham to Somerton, which thankfully was not implemented. A report on this proposed transfer reads;

"We recommend that the principle of relocating Burnham facilities to Somerton is abandoned, but consideration given to the utilisation of omni-directional antenna receiving facilities for all services and relocation of receivers to Burnham. The requirement for Somerton as an operational site would then cease and the requirement for upgrading the power distribution, alarm systems and need for maintenance support at Somerton would then cease. The site could be moth-balled or disposed of as appropriate".

The report also mentions the problems of cost, timescales, and disruption to services, all of which would be unacceptable. In the event, the use of Somerton receiving aerials and the radio receivers was maintained until the closure of the service in April 2000.

R/O Steve Allison at one of the W/T consoles in the early 1990s, showing the computer screen which replaced the 'Trend' teleprinters during the 1980s.

Use of the radio telex 'databank' was increased, with vessels able to interrogate the database to obtain weather, weekly football results (a service agreed between BT and the BBC), order flowers for loved ones, select greetings telegram designs, check the latest BBC World Service Schedule, obtain radio propagation forecasts and other useful information. Ships could even obtain the weekly national lottery numbers by entering LOTTERY INF+ as well as other information added on a regular basis.

The station continued to contract in size, and the operating area was halved when BT's International Customer Service Centre (ICSC) moved into the main building in 1990. All operating consoles were then concentrated in one small area, and the landline room also reduced in size to make way for more offices. Unfortunately, this meant that access to the operating area was via a circuitous route around the outside corridors of the building, as the new occupants of the building would not allow the R/Os through their section. A few staff from the radio station actually obtained jobs within the ICSC when it became clear that their future was in doubt.

BT were keen to utilise the building and as more groups moved in, the radio station operating area became smaller year by year; former office areas were also taken over by other groups and staff numbers increasingly declined.

The W/T area in the early-mid 1990s. Only 6 consoles remain, and the location of the former Overseer's 'Wendy House' can be seen by the floor markings.

A further attempt to prolong the radiotelegram service was the re-introduction of the 'after office hours' delivery service, where for a small charge, telegrams would be telephoned to a designated company representative in addition to the normal delivery by telex. Many companies actually took this option up, and it proved popular until more ships became equipped with Inmarsat equipment.

Companies who preferred to use foreign coast stations for their traffic were also contacted by Portishead advising them of the cost savings involved in advising their vessels to contact Portishead directly.

The station regularly received reception reports from short-wave radio listeners from all over the world, and although the station did not encourage listening to the content of messages and R/T calls, all reports were acknowledged with a 'QSL card' to confirm receipt. Word soon got around amongst the radio listening fraternity and dozens of reports were received each month.

An interesting link was established by the station in 1993. A bomb disposal unit working in Somalia for the United Nations known as 'Rimfire' used the radiotelephone service regularly to phone home. On their return, the organisation complimented the station for their "hard work and professionalism".

GKA

Portishead Radio
BTI Radio Station
Highbridge
Somerset
(TA9 3JY)

Radio Stations
Portishead · Wick · Stonehaven ·
Cullercoats · Humber · North Foreland ·
Niton · Lands End · Ilfracombe ·
Anglesey · Portpatrick ·
Satellite Earth Station
Goonhilly

Centres of Excellence for Maritime Communications

QSL Card sent to radio enthusiasts on receipt of reception reports.

That same year, Ralph Featherstone, skipper of the yacht 'Valiant Lady' completed a three-year solo voyage around the world. During this voyage he used Portishead for his radiotelephone calls home and thanked the staff "for their patience and help in keeping me in touch with home during my three years away".

The 11th February 1995 was significant in one historical way; this was when the last ever radiotelegram from a British vessel fitted only with W/T equipment was received at Portishead. There were a few subsequent occasions when British ships would call on W/T but the vessels were equipped with satellite or radio telex equipment (probably unserviceable on the day). The vessel in question was the *'Estireno/MRCK8'*, who sent a short ETA message. Never again would a British ship equipped only with W/T equipment contact Portishead. The end of an era.

Many deep-sea fishing fleets, notably from Morocco and South Korea, made numerous radiotelephone calls via Portishead, although sadly very few paid their bills; this resulted in many of these vessels being placed on the 'stop list', meaning only distress and safety calls would be handled.

September 1995 saw the celebration of the 75th anniversary of the long range maritime radio service in the UK. Numerous events were held, including a large staff reunion, TV and newspaper coverage, commemorative first-day stamp covers and the issuing of special ties to all staff. Richard Branson, pilot of the *'Virgin Global Challenger'* Balloon and the *'Virgin Atlantic Challenger'* speedboat, was happy to congratulate the station:

"In 1987 when Per Lindstrand and I crossed the Atlantic by balloon and subsequently in 1991 when we crossed the Pacific, Portishead was a godsend. They kept us in touch via our HF radio and their landline links. It was through that we found that the Gulf War had started and they helped save our lives on more than one occasion. We look forward to being in contact this winter when we attempt the first non-stop circumnavigation of the globe in the Virgin Global Challenger balloon".

Sir Ranulph Fiennes frequently used the service over 20 years during his transpolar expeditions as satellite communications were unavailable in the Polar Regions. He and his wife Ginnie, who also his base leader, explained how they became friends with some of the Portishead staff without actually meeting them by mutual contact during long expeditions. Sir Ranulph felt it was thanks largely due to Portishead that he was able to complete the first-ever journey around the Earth's circumpolar axis, crossing both North and South Poles.

"The operators always went out of their way to help Ginnie, advise her and wade through interference and weak frequencies rather than give up on the communication. Overall Portishead has been incredible in both kindness and efficiency. We welcome the chance to commend 75 years of service and thank them for their help during all our expeditions" Sir Ranulph said.

Lisa Clayton, who circumnavigated the globe single-handed in the *'Spirit of Birmingham'* yacht, felt indebted to the Portishead staff. "Their support and personal attention, including knowledgeable advice, helped me cope during some very trying times. They were brilliant," she said.

Commemorative First Day Cover from 1995.

Press reports from 1997 advised that the future of the station was uncertain, to say the least, as one article stated;

"Portishead Radio, which operates from Burnham-on-Sea, Somerset, is now one of the world's largest remaining stations receiving and transmitting ship-to-shore communications in Morse code. However the company, part of British Telecom International, also faces the prospect of life without Morse. It has reduced its staff from 300 radio operators 15 years ago to just 30 as it now receives only 100 messages a day, compared with 1,000 in the 1970s. Operations Manager Peter Boast said:

"The writing is on the wall - it is just a question of when. Everyone would be sad if it's the end as it's a good communications system, but it has almost run its course." A BT spokeswoman said there were "no immediate plans" to close Portishead Radio but acknowledged that Morse code was not likely to survive beyond 1999."

The end of 1997 saw plans to end the 500 kHz distress watch around the UK coast along with the commercial Morse service. The short-range stations broadcast the following message:

FROM 010000Z JANUARY 1998 THE COASTGUARD AGENCY NO LONGER REQUIRES BRITISH TELECOM MARITIME RADIO SERVICES TO MAINTAIN 500 KHZ DISTRESS WATCH.

AT THE SAME TIME, BRITISH TELECOM MF COAST RADIO STATIONS WILL CEASE ALL MORSE COMMERCIAL SERVICES.

THE MORSE SERVICES WILL CONTINUE AS NORMAL VIA PORTISHEADRADIO/GKA.

MF COAST STATION STAFF SEND BEST WISHES TO ALL RADIO OPERATORS, PAST AND PRESENT, WHO USED BRITAIN'S MF MORSE SERVICES DURING THE PAST 89 YEARS =

BRITISH TELECOM MARITIME RADIO SERVICES 01 NOVEMBER 1997 +

Further writing was on the wall from July 4th, 1998, when the HF W/T service of Athens Radio/SVA closed. Athens was one of the largest W/T stations in Europe, due to the number of Greek-flag vessels with suitable equipment, and its closure was to signal the decline of HF W/T services throughout Europe.

Many newspapers jumped on the bandwagon, with numerous articles featuring the demise of commercial Morse code. Don Mulholland, former Station Manager was quoted as saying:

"It will be a sad day. Morse was very romantic. One minute you could be talking to vessel in the Bristol Channel and the next to a cargo ship in some hot, exotic place on the other side of the world."

Another Station Manager, Ernie Croskell, was of the same opinion. In an interview for local television, he quoted that "there was no better or efficient means of communication than 2 damn good Morse code operators working together". I am sure most of the staff agreed.

The staff restaurant was closed in May 1998, despite vociferous protests from staff and unions. In fact, the protests made the local press and television, but all to no avail. The nearest BT Staff restaurant was now located in Bristol, some 30 miles away.

Protests outside of the station entrance against the closure of the staff restaurant.

The welfare block, home to the staff bar and lounge area was closed in March 1999, again to great sadness. At its busiest in the early 1980s, there was a major function each month, with the staff Christmas Party and New Year celebrations being a highlight. Combined with the evening darts and skittle evenings, plus quiz nights, private functions and similar events, the club became a thriving business, with the bar open each lunchtime and evening. It became a regular haunt for those R/O's on the evening shift, whose 15-minute break was well spent with a quick pint or two before returning to work.

Friday, March 19[th,] 1999 saw NO SHIPS contact Portishead Radio on W/T for the first time in its history. Traffic figures of 5 or 6 a day were not unusual at this time, but a nil return was a very poignant occasion. By now, there were only 2 W/T consoles in operation at the station, and these could also be used as extra radiotelephony consoles if required.

Many attempts to bring new services to the station were tried. British Airways considered closing their Speedbird HF radio network to allow BT to take over their system, but this was rejected by BT. Instead, use of the Rugby transmitters was offered to British Airways. Globe Wireless made enquiries about using the station as their European hub for HF data communications, but again this was rejected. There was even the opportunity for Portishead to become the European HF GMDSS control station, but this role eventually went to Lyngby Radio in Denmark.

The station took over BT's Telex Bureau services for a short time, handling messages telephoned or faxed into the station for telex delivery world-wide. This was a highly manually-intensive service and not at all popular with the staff.

As GMDSS became fully operational in 1999, the end of the coast radio station network was imminent. Many of the services were ceased as indicated by the following broadcast:

CQ DE GKB = QSX 4 8 12 AND 16 MHZ = EFFECTIVE 31/2359Z JANUARY 1999 PORTISHEAD RADIO WILL CEASE NAV WARNING AND WX BROADCAST ON W/T. WEATHER BROADCAST WILL CONTINUE AS NORMAL ON RADIOTELEX AND RADIO TELEPHONE. W/T WILL CONTINUE TO BE AVAILABLE FOR TRAFFIC AFTER THIS DATE. = PORTISHEAD RADIO 15/1200Z JANUARY 1999 +

However, BT seemed reluctant to make the final decision with regard to the closure of Portishead as this broadcast from August 1999 confirms:

CQ DE GKE 1500190899 R13518=

SHIPS' NOTE: THE DECISION ON THE CLOSURE DATE FOR PORTISHEAD RADIO HAS BEEN POSTPONED. CLOSURE WILL NOT NOW TAKE PLACE UNTIL FRIDAY 17 SEPTEMBER AT THE VERY EARLIEST=

PORTISHEAD RADIO 191200Z AUG 99

It was obvious that BT had decided that the time had come to close all the remaining stations (including Portishead), but the actual date of closure was still the subject of negotiation. However, it was decided to prolong the service into 2000 in case any problems with the 'millennium bug' were encountered and HF radio services would be required. As we all know, 2000 arrived without any issues, and it was formally decided to close the station on 30th April.

The closure of the station was announced by way of a broadcast during April 2000;

CQ CQ CQ DE GKB GKB GKB = REGRET TO ANNOUNCE THE CLOSURE OF GKB AND ALL UK VHF SERVICES AT 1200Z SUNDAY 30 APRIL 2000. MF STATIONS WILL CLOSE AT 1200Z ON FRIDAY 30 JUNE 2000. WE SEND OUR THANKS AND BEST WISHES TO THE MARITIME COMMUNITY, WHICH WE HAVE SERVED FOR OVER 90 YEARS = MARITIME RADIO SERVICES, LONDON, 30 MARCH 2000 +

On Saturday 29th April 2000 from 0700z to 1900z a special once-only and c.w. only Amateur Radio event took place with cross-band communications between radio amateur stations and Portishead Radio.

With Portishead Radio using its multi-kilowatt transmitters at Rugby and its remote receiving array at Somerton, the station's Radio Officers at Burnham gave Radio Amateurs, sitting in the comfort of their own radio shacks, the chance to communicate directly with GKA before the station's transmitters fell silent for the final time.

Three stations operated at any one time, subject to the commercial requirements of the station, and special efforts were made to beam towards Commonwealth countries at appropriate times.

BT appointed the Radio Officers Association to handle the amateur side of this operation and the liaison officer was David Barlow G3PLE. All contacts received a QSL card via the RSGB bureau.

The event was marked by a broadcast from Portishead Radio;

CQ DE GKB 2/4/5/6/7 =

THIS UNIQUE MARITIME/AMATEUR CROSS BAND EVENT IS NOW AT ITS END. THE RADIO OFFICERS ASSOCIATION WISH

TO THANK BT MARITIME SERVICES AND THE RADIO COMMUNICATIONS AGENCY FOR MAKING IT POSSIBLE. QSL ALL CONTACTS VIA BURO. NO INCOMING CARDS PLEASE. ONLY S.A.E TO PO BOX 50 HELSTON TQ12 7YQ ENGLAND.

LET'S REMEMBER ALL THOSE WHO DID NOT MAKE IT BACK TO PORT BOTH IN WAR AND PEACE TIME. THEIR LOG BOOKS CAN NOW BE CLOSED AND THEY CAN GO OFF WATCH AS WILL PORTISHEAD RADIO TMW. TO ALL OF THE AMATEURS, MNI TKS NW QTP & CL SEE U TMW FOR LAST TIME =

TU AND GOOD BYE + VA

In the days leading up to the station closure, the following broadcast was made;

VVV VVV VVV DE GKB CQ DE GKB = QSX 4 8 12 16 MHZ +

CQ DE GKB = THE LAST BROADCAST FROM PORTISHEAD RADIO WILL BE ON THE 30TH FIRSTLY ON R/T 4384. 8764 AND 13146 THEN ON R/T 17245, 19755 AND 22417. THIS WILL FOLLOWED BY THE WT BROADCAST ON GKB2/4/6/7. THE RTT BROADCAST WILL BE AUTOMATIC AT 1201Z + QRU + DE GKB +

Sunday, April 30[th,] 2000. A poignant and indeed sad day. This was to be the last day of operation of Portishead Radio.

As the morning shift arrived, it became clear that this was not going to be an ordinary day. Ships called in not to exchange traffic but simply to speak to the station for the last time. Many staff members came to the station to witness the last hours. Retired staff members arrived during the morning for one last look around. Tears were shed. Local and national media arrived with cameras and recorders to retain the last broadcast for posterity. One of the Radio Officers on duty decided to make his own broadcast;

CQ DE GKB =

THIS MESSAGE IS FROM AN R/O WHO HAS BEEN WORKING INTO AND FROM GKB SINCE 1963 =

KRS AND 73S =

JOHN HOCKING +

Over 200 people crammed into the operating area as the clock slowly moved towards 1200 GMT, after which the station would be no more. The old Post Office voice announcement marking the R/T Channels was resurrected for the last time. As 1200 GMT arrived, Radio Officer Larry

Summers prepared to make the last R/T transmission, immediately followed by Mike Pearson on W/T.

John Hocking, whose personal broadcast was heard prior to the final transmission.

The broadcast read as follows:

CQ CQ CQ DE GKB2/4/5/6 =

THIS IS THE LAST BROADCAST FROM PORTISHEAD RADIO. FOR 81 YEARS WE HAVE SERVED THE MARITIME COMMUNITY. WE SAY THANK YOU TO ALL THOSE WHO HAVE SUPPORTED AND USED OUR STATION.

WE PAY TRIBUTE TO MARCONI WHO MADE IT ALL POSSIBLE. HIS FIRST TRANSMISSIONS ACROSS WATER WERE MADE FROM NEARBY HERE AND SO STARTED THE RADIO ERA. WE ARE PROUD TO HAVE BEEN PART OF THAT ERA.

AS THIS HISTORIC TIME IN THE COMMERCIAL MESSAGING WORLD COMES TO A CLOSE THE MANAGER AND RADIO OFFICERS WISH YOU FAREWELL FROM PORTISHEAD RADIO/GKB + VA

The transmission was also broadcast automatically by radio telex at 1201 GMT.

And that was it. To a round of applause, the staff dispersed to the welfare club for further celebration and reminiscing. The receivers were switched off and the engineers moved in with indecent haste to commence the dismantling of the consoles.

R/O Mike Pearson about to send the last W/T transmission from the station.

Rugby 'B' transmitting station closed at the same time, bringing an end to the maritime service from that famous site.

I was on duty that morning working for the BT Inmarsat service, so I made my way back to my office. The telephone rang and I answered it as no-one else was around. It was a lady wishing to book an R/T call to her husband on board one of the offshore rigs in West Africa. Sadly I had to inform her she was just a few minutes too late.

The 'wake' in the welfare club (specially re-opened for the occasion) lasted a good few hours, by which time many of the consoles were already in a dismantled state. Most of the hardware ended up in skips outside of the station building, as well as most of the consoles.

The buildings continued to be used by other BT groups (mainly the International Customer Service Centre (ICSC)) until the decision was made to sell the land in 2007. BT's Airwave service was based at the station, and BT's Satellite services employed a small team of ex-Radio Officers who were housed in a small office in the former landline area. This team was responsible for the service activation of UK-registered Inmarsat terminals, the administration of BT's C-Sat services, Inmarsat fraud investigation, management of the 'stop list' and general customer service work.

A service called 'Webtrack' was introduced, which was a system used by both maritime and transport fleet customers to monitor the location of their vessels or lorries using Inmarsat-C and GPS tracking. In addition a service called 'Satmail' was established where customers could use their e-mail systems to send messages to ships over satellite for e-mail, fax or telex delivery using a mailbox at Goonhilly Land Earth Station. Both of the above were administered from the office.

The office was open from 0800-2300 daily, with queries received during the night being passed to Goonhilly Land Earth Station for action.

For a short time, the office handled Iridium terminal activation, although this service ceased when the operators filed for Chapter 77 bankruptcy. New owners revived the system a few years later and it has recently become a fully GMDSS-compliant system.

Plans to vacate the building were explored, with options of moving the team to the Burnham-on-Sea telephone exchange and then the Weston-super-Mare telephone exchange explored. However, by 2001, BT had sold its Aeronautical and Maritime business to Stratos Global Communications of Canada, who did not require the use of the Highbridge site. All remaining BT Satellite Services staff were 'redeployed' with some moving to other BT departments and others taking a redundancy package.

Some redeployees were instructed to attend work, with temporary jobs being assigned by BT on an 'ad hoc' basis, mostly mundane and uninteresting administration tasks. One such task was to advise BT staff across the country of their projected redundancy payments should they choose to leave the company.

A few took employment at the UK Hydrographic Office in Taunton, working in the Admiralty List of Radio Signals department. Only the ICSC remained at the station after this time, before they themselves were relocated to the Weston-super-Mare telephone exchange on Friday, March 11[th,] 2005, leaving the building totally uninhabited.

Sadly this gave rise to a significant degree of vandalism and destruction. The site was regularly broken into and all internal fittings were destroyed or damaged, and many external doors and windows damaged beyond repair. The station was recognised by the local history group as part of the 'Highbridge Heritage Walk' and a locally-produced blue plaque commemorating the station was mounted on the entrance building. Unfortunately, the plaque was stolen and the station eventually removed from the Heritage Walk Guide

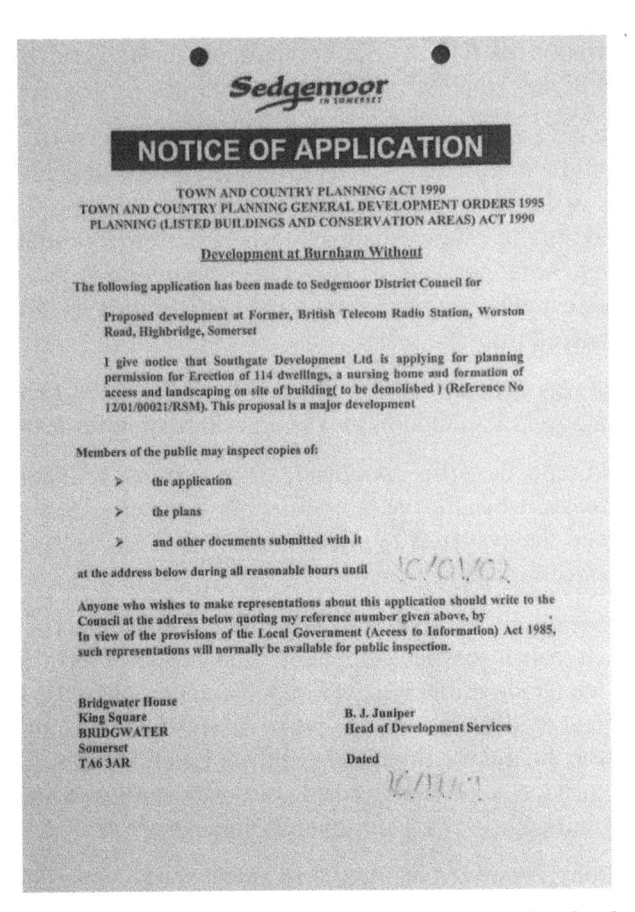

> ### Sedgemoor
> IN SOMERSET
>
> ## NOTICE OF APPLICATION
>
> TOWN AND COUNTRY PLANNING ACT 1990
> TOWN AND COUNTRY PLANNING GENERAL DEVELOPMENT ORDERS 1995
> PLANNING (LISTED BUILDINGS AND CONSERVATION AREAS) ACT 1990
>
> #### Development at Burnham Without
>
> The following application has been made to Sedgemoor District Council for
>
> Proposed development at Former, British Telecom Radio Station, Worston Road, Highbridge, Somerset
>
> I give notice that Southgate Development Ltd is applying for planning permission for Erection of 114 dwellings, a nursing home and formation of access and landscaping on site of building(to be demolished) (Reference No 12/01/00021/RSM). This proposal is a major development
>
> Members of the public may inspect copies of:
>
> - the application
> - the plans
> - and other documents submitted with it
>
> at the address below during all reasonable hours until 10/01/02
>
> Anyone who wishes to make representations about this application should write to the Council at the address below quoting my reference number given above, by .
> In view of the provisions of the Local Government (Access to Information) Act 1985, such representations will normally be available for public inspection.
>
> Bridgwater House
> King Square
> BRIDGWATER
> Somerset
> TA6 3AR
>
> B. J. Juniper
> Head of Development Services
>
> Dated

The first application to demolish the Highbridge site for housing, 2002. The proposed nursing home was never built.

The buildings were all demolished in 2007 to make way for the 'Mulholland Park' housing estate, named after the former station manager Don Mulholland and his father Robert. The road names were unimaginatively named after some electronic/electrical pioneers such as Marconi, Tesla, and Susini, despite a suggestion from former staff members to remember the coast radio stations of the UK with names such as Niton Way, Cullercoats Walk, and Portishead Drive, etc.

The demolition of the station also put to an end the rumours of the old building being haunted by the legendary 'Red Eyes'. Many a staff member would settle down for a nap on the night shift in the old building but would wake up to a ghostly apparition in front of them. No evidence has been found of any staff member passing away within the building, and it is fair to assume that 'Red Eyes' can be described as a myth. However some staff may remain to be convinced, some of the view that the 'ghost' is that of ex-R/O and Doorman Frank Davis, who allegedly took his own life.

The boarded-up entrance building looking sad prior to demolition.

In 2008, the band 'Portishead' released their 'Third' Album, which featured a photograph of the Highbridge receiving station on the inner sleeve of the CD. It made number 2 in the UK album charts, and following the release of Mike Batt's 'Waves' album in 1980, was the second album to feature the station.

At the time of writing, nothing exists on or near the site to show that the most famous long-range maritime radio station in the world was once there. Plans for a memorial and a suitable plaque have obtained support from ex-staff, the local and county councils and the area's Member of Parliament, but as yet nothing has been decided. However, it is hoped that some sort of memorial will appear on the site in the near future.

The site these days consists of numerous houses of varying styles and designs. There is a small green area at the front of the estate which would have been to the right of the station entrance. However, there are still parts of the station left; in the fields surrounding the housing estate, there are a few sections of concrete on which the aerial masts were mounted. Otherwise nothing remains.

The last moments of the station entrance building, 2007.

Some ex-staff who are radio amateurs regularly participate in Maritime Radio Day, held in April each year, and special call signs GB0GKA, GB0GKB and GB0GKC (originally activated for the 80th anniversary of Portishead Radio in 2018) are activated for this event and also used on special occasions when required. In 2010, a special call sign of GB10GKA was activated to commemorate 10 years since the closure of the station, again operated by a team of ex-staff. Call-sign GB0GKU was allocated to celebrate the 100th anniversary of the service in 2020.

A website at www.portisheadradio.co.uk is maintained which contains a wealth of information about Portishead Radio and its history, together with audio and visual files concerning the station.

The construction of 'Mulholland Park', c. 2009.

2020 view of Mulholland Park from the same location.

The remaining staff continue to meet up for meals and talk about the station on a regular basis, to exchange stories and to keep up with the whereabouts of former colleagues. Formal reunions take place every few years. At least we have our memories of what was once a wonderful place to work. A sad and ignominious end to the world's largest and busiest maritime radio station.

Portishead Radio. We salute you.

CHAPTER EIGHT

AIRCRAFT, YACHTS AND FIXED STATIONS

Portishead Radio had been involved in communication with aircraft since the 1930s, and indeed during the Second World War; but it wasn't until 1983 that a dedicated aeronautical service was launched. Some aircraft, fitted with suitable equipment, had been able to use Portishead's maritime R/T frequencies, but these were few and far between.

Aeronautical R/T section showing the SELCALL units mounted on top of each console.

So it was, in early 1983 that the new dedicated HF Aeronautical Radio service was trialled, with test calls made from aircraft owned by Dan-Air and British Island Airways. It took the Portishead staff some time to come to terms with aeronautical language, and many pilots were taken aback by being referred to as 'Old Man', a term regularly used in the maritime radio service. As the staff became more used to the slick operating techniques utilised by pilots and crew, other airlines became interested in the service.

Aircraft now had the opportunity to call Portishead to request a telephone call to virtually any number world-wide, with the cost of their calls being charged to a nominated U.K. telephone number, to any address world-wide, or to a credit card. To prevent credit card information being given over the air, an account scheme was devised whereby a 5-digit 'PIN' number was allocated to each account holder, ensuring that sensitive credit card information would not be disclosed.

The major difference between maritime and aeronautical calls was that the latter operated on one frequency only (simplex), thereby precluding both parties from speaking at the same time. There were some initial equipment problems, mainly to do with feedback, but these were rectified by judicious use of commercially available terminal units which utilised voice-operating suppression units. Quick-tune transmitter control units were installed which enabled the Portishead R/O to bring up a transmitter on any frequency in a matter of seconds.

The results of these initial tests proved encouraging, and it was decided in 1984 to test the market with visits to many airlines by representatives of Portishead Radio; these again resulted in a positive response from both large and small airlines, and in the summer of 1985 the Portishead Radio Aeronautical Service was formally launched.

Dedicated operating consoles were constructed from surplus maritime consoles, with a 24-hour watch being kept on the designated channels allocated (existing point-to-point frequencies). As a promotional incentive, first-time callers were given a free call to see for themselves how the system operated, a move which proved highly successful.

Portishead joined the SITA network, which meant that aircraft could tender messages at any time for relay to their operations or engineering departments; in addition, Portishead could obtain weather information from virtually any location on request from aircraft. The weather information was supplied in code, so a great deal of time was spent by staff learning what the figures actually meant; however in true R/O fashion, such codes were quickly learnt and many staff members became proficient in understanding and transmitting the information.

As an aid to pilots, regular propagation charts were issued, advising the best frequencies to use to contact Portishead at any time from a selection of popular locations. These proved so popular that over 1,000 were distributed every couple of months.

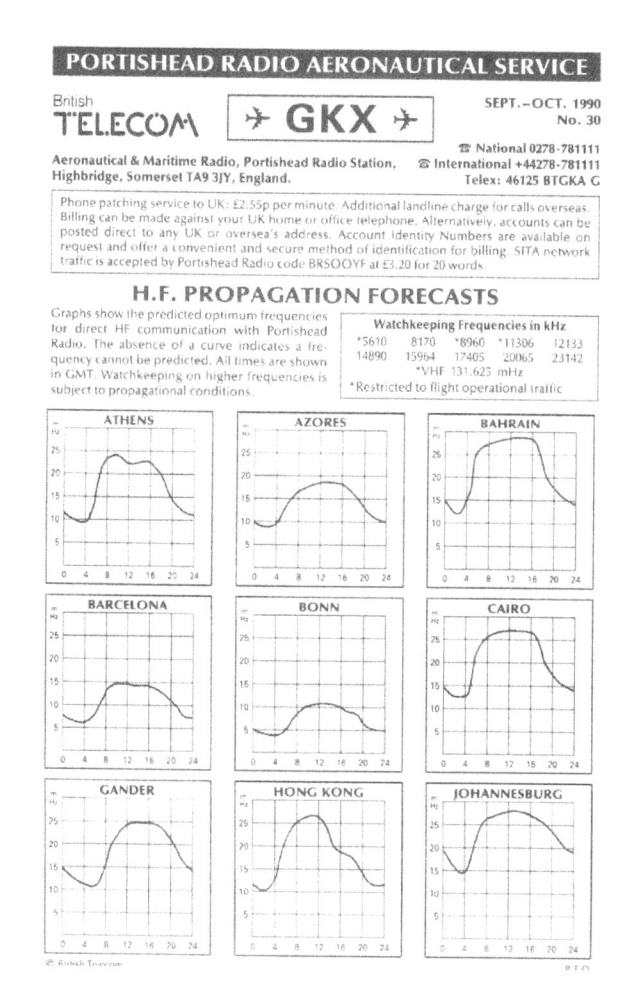

Example of the HF aeronautical radio propagation chart as issued to pilots.

A promotional leaflet issued by BT stated;

"You can now make telephone calls from your aircraft to anywhere in the world. British Telecom's high-frequency radio station at Portishead is offering a brand new service to act as your UK and International telephone operator. To use this service all you have to do is give Portishead the number you require and we'll put you through, 24 hours a day, 7 days a week.

With a call charge based on a one-minute minimum (rather than the three minutes charged by our nearest competitor), we offer a far cheaper service than our rival systems. And another overwhelming advantage is that Portishead links into over 160 countries on the International Direct Dialling network, giving you a truly world-wide service. Portishead has the latest generation of aeronautical communications equipment, and the station is manned by over 150 fully trained Radio Officers who are ready to offer you the best possible service.

Once you're through to Portishead, just give your call sign, flight number, required telephone number and billing address. You'll be charged only £1.62 per minute for UK calls and £1.56 per minute for calls abroad (plus the normal tariff from the UK to the specified country). Accounts are forwarded every quarter. And if you have a UK telephone number, you can have the call charged to that number, from anywhere in the world, even if you're calling from Portugal from above the South Pole.

Tune your transceiver to our frequencies and your first call will be absolutely free of charge. It's a great way to test the system.

Portishead responds to aircraft carrying both simplex and duplex HF equipment. What's more, we can easily link you into the SITA network for a charge of only £3.00 for up to 20 words. You can also be alerted to receive calls, via Portishead's SELCALL service. So we can really keep you in touch. Our automatic telex service is also available for specially equipped planes.

At Portishead, we maintain a continuous monitor of the following frequencies (kHz):

4807 8170 12133 16370 21765

Additionally, the following frequencies may be used on request (kHz):

4810 8185 12168 17405 18210 19510

The latest free propagation charts are available on request".

In addition, credit-card sized plastic cards were issued with details of frequencies and contact details, and provided free of charge to all users of the aeronautical service.

BT Radio Station – Portishead Radio
Aeronautical and Gateway Service

The following frequencies (kHz - USB) are monitored by Portishead Radio:

4807	5610*	6634*	8170
8960*	10291	11306*	12133
14890	15964	16273	17335
18210	19510	20065	23142

*restricted to aircraft operational traffic only.
A VHF channel on 131.625 MHz is available for aircraft operational traffic only.

Services provided

- Selcall checks (to aircraft) and radio checks free of charge
- Medical advice offered free of charge
- Radio propagation charts regularly distributed on request to customers
- One minute minimum charging
- Account numbers available to allow billing by credit card or mail
- Calls to UK may be billed to the destination telephone number
- SITA messages accepted from aircraft for onward transmission
- Airport weather information available on request (chargeable)
- For further information ring Portishead Radio Freefone 0800 37 83 89

BT *The 'key' to your communications needs*

The credit card-sized information cards sent to pilots.

As news of the service spread, Eastern Airlines of the USA decided to use Portishead as its main European communications station; a decision which brought in extra business and associated revenue. It also meant that Portishead could now operate on the aeronautical 'R' band frequencies instead of the more obscure point-to-point channels. At the insistence of Eastern Airlines, Portishead adopted the working name of 'UK Radio Flight Support' for their customers as it was felt that Portishead was too obscure a name to remain familiar to American pilots. However, the ultimate demise of Eastern Airlines curtailed this operation although some of Eastern's customers continued to use the services.

In addition, most UK-based charter airlines used the station, as well as numerous cargo airlines operating out of Africa. Preferential rates were offered to regular users of the service, and it was often reassuring to hear pilots using Portishead rather than the competing stations at Bern and Stockholm.

To confirm Portishead's status as a major aeronautical service, a commemorative plaque was awarded to 'The World's fastest-growing aeronautical station' in 1988.

Portishead handled some notable 'firsts' on the aeronautical frequencies; Richard Branson's Transatlantic balloon crossings were probably the most memorable, as all voice communication was handled by the Portishead operators, contrary to press reports which stated the calls were made by satellite! Numerous fund-raising ventures have been co-ordinated by Portishead over the years, especially during major events such as 'Children in Need' and 'Red Nose Day'.

The service proved popular with many pilots, and some have some wonderful memories of the service;

One captain, heading south from Manchester in a Boeing 720, flew over his house near Leek, saw his son had turned on about 4kW of floodlighting to play football in the farmyard - thought he was safe to waste the electricity now Dad had gone to work - wrong! The poor lad answered the phone to Portishead and a severe telling-off...

One story involves Steve Fell, one of the Portishead Radio R/Os. It was summer, nice and quiet, Steve on Aero R/T as he usually was on nights, the door was open into the field as it was a warm early morning. He had a contact from a Britannia aircraft who had a quick call to Ops then, as he wasn't busy either, started chatting with him as to where exactly Portishead Radio was situated, as he knew it wasn't Portishead. Steve explained he was in Highbridge, and the pilot said he thought as much, and was lining up to land at Cardiff, so should pass over Highbridge shortly, if Steve wanted to nip outside. It was quiet on R/T so Steve wandered out into the field. Not a sight nor sound of anything in the air, so after a while he went back in. Whereupon the plane made contact again. Steve said he hadn't seen anything, and the pilot said: "Just coming up to you now, keeping nice and low, then I'm turning to line up for landing at Cardiff." Steve went outside again and looked up - nothing - then looked behind him and DOWN a bit - whereupon the sky was blotted out by a huge shape. The aircraft was on minimum power and Steve hadn't heard the approach. Apparently, if the old aerials had still been in place he wouldn't have made it - he was that low. Steve said he was shaking after this - and the pilot could have got in serious trouble had it been reported.

Many Portishead staff were invited to the flight deck on holiday flights by grateful pilots and allowed to use the HF radio en route. I recall a Britannia flight to Luxor when I was on the flight deck for over 3 hours, even managing a contact with Portishead on 16 MHz, leaving my good lady fuming by herself watching the in-flight films. I politely declined the invitation to go 'up top' on the return flight.

When the GKA football team flew to Switzerland to attend the annual European football tournament, the flight was booked with Swissair – a company that exclusively used Bern Radio for their HF radiotelephone contacts. However, this did not stop the Station Manager of the time (who came with the team) to try to give the crew free copies of propagation charts and frequency cards. He gave the pack of paperwork to the stewardess to pass to the pilot and co-pilot, but was soon dismayed when the stewardess returned with the paperwork and the immortal words "The Captain says no".

It was soon realised that land-based stations such as those owned by relief agencies and charities could also make use of the facilities; Organisations such as Oxfam, MSF (Médecins Sans Frontieres), Save the Children, British Direct Aid and the Red Cross have all used the service for vital and morale-boosting calls home. Calls from these 'land-based' stations constituted the Gateway service, kept separate from the aeronautical service although the same R/T frequencies were shared. In order for any messages to these stations to be handled, locally-issued 'dummy' call signs beginning 98 or 99 were issued (98OXF or 98MSF for example), and for radio telex, a locally-issued SELCALL number beginning 76xxx was issued.

During both Gulf Wars and the Balkan conflict during the early 1990s, service personnel could call home using the aeronautical simplex frequencies, the cost of calls being reverse-charged to the destination number if in the UK. Some USA networks also accepted reverse-charge calls, but the mark-up charged by these networks made the cost of these calls prohibitive. Calls from USA personnel stopped virtually overnight as the bills began to be received.

It was not unusual, especially from the Balkans, to receive radio calls from tank-based personnel, requiring a call back home. Such calls were excellent for morale and proved extremely popular.

A promotional leaflet describing the Gateway service stated;

"Gateway is based on voice and telex communication.

Telephone: Your Gateway link to Portishead Radio gives you operator-controlled access to telephones in the UK and around the world.

Telex: Auto-dial, auto-correction and auto-receive service for store-and-forward and conversational modes, with or without operator service.

Gateway also offers a number of message relay services which extend the range of the automatic radio telex service.

Phonetex: Anyone in the United Kingdom can call Portishead Radio by telephone and have a message relayed to the designated terminal by telex.

Radiopaging and Voicebank: Pager alerts and Voicebank messages can be placed by Portishead Radio on receipt of telex instructions from the remote terminal.

Radiotelex letters and telegrams: A message telexed to Portishead Radio may be posted to any address in the world or delivered as a telegram by the best available means.

Semi-fax: Messages telexed to Portishead Radio may be sent to any facsimile terminal or group of terminals world-wide.

Customer courses: Portishead Radio also runs a one-day training course for radio telex users as well as providing back-up services whenever required".

The leaflet also gave details of the organisations likely to benefit by using the Gateway service, such as relief agencies, industry, embassies or consulates and shipping agents. The station could also put the customer in touch with suitable equipment manufacturers or dealerships in order to expedite the commissioning of such stations, although no preference or recommendation could be shown.

The aeronautical service was, however, much more lucrative, and new equipment (including the use of rotatable log periodic aerials and improved radiotelephone/landline terminals) was obtained in order to enhance the service.

Unfortunately, as the recession hit the aero industry in the early 1990s, more and more carriers fell by the wayside, and some existing airlines decided to set up their own communication systems. This naturally caused some concern at Portishead, but the increased power and range of Portishead's transmitters served as an excellent back-up for airlines whose own systems were not flexible enough for long range communications. In fact, Britannia Airways decided to scrap their own service and use Portishead exclusively, with others preferring to use the station rather than their own basic systems.

Regular contacts were established with the M.A.F.F. fishery protection aircraft around the United Kingdom, who maintained hourly schedules when on operation.

The actual consoles (of which there were two) used for the aeronautical service each monitored 9 separate channels on loudspeakers mounted on top of each unit. Two Racal receivers were installed on each console, together with two Marconi quick-tune transmitter (QT3) control panels, in turn, linked to the transmitters themselves at Rugby. A directional aerial system using rhombic receiving aerials located at Somerton ensured the best possible signal could be received. Rotating Log Periodic (RLP) transmitter aerials were brought into use later on, but these proved to be a little temperamental and tended to 'stick' on occasions. However, the initial problems were quickly overcome, and they continued to provide an excellent service until the station closed.

Motorola 'SELCALL' units were also available, giving the Portishead R/O the facility to call any aircraft known to be listening on specified frequencies 'on demand'.

It was possible for 4 circuits to be running simultaneously (2 per console), with the left-hand console operating on frequencies below 13 MHz, and the second one between 13 and 25 MHz. However, both consoles had the facility to operate on any frequency required. A VHF Channel was available on both consoles for use by aircraft over the South and West of England.

Aircraft would be able to avail themselves of free SELCALL checks, weather information and SITA messaging. In addition, should an aircraft have required medical assistance, calls would be patched through to the duty doctor at the Royal Naval Hospital in Gosport for advice.

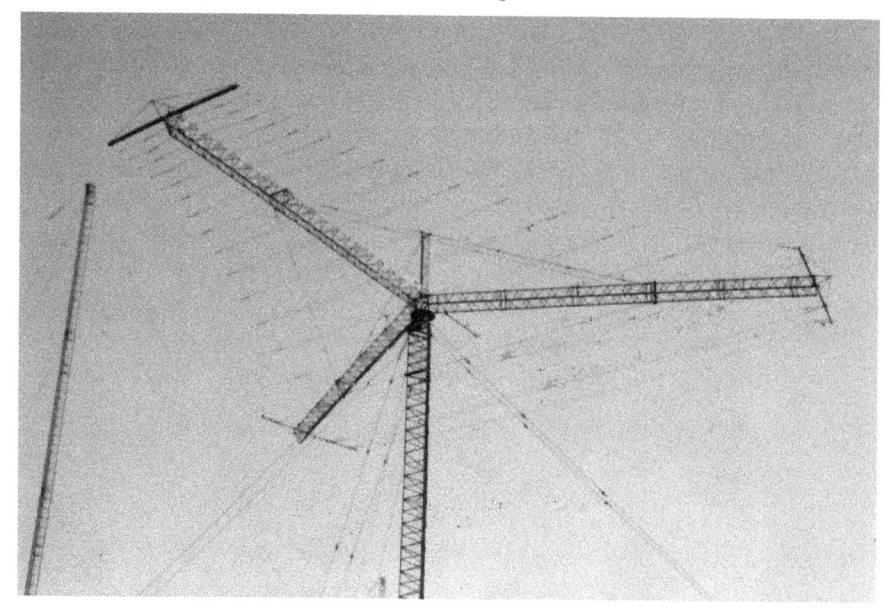

Rotating Log Periodic Aerial at Rugby.

During 1987, BTI introduced the Skyphone service, a satellite-based system providing access to the world's telephone service. Additional services such as weather, share prices, and hotel bookings would use the aeronautical department at Portishead, who would access various databases to obtain such information. However, it was rarely used in the early days, and despite increased awareness, the service never really established itself at Portishead.

Unfortunately, towards the end of the 1990s, air crews could avail themselves of mobile telephone technology whilst 'on the ground' and with the advent of satellite communications, it was clear that the aeronautical service was on borrowed time. However, enough traffic was maintained to ensure the service continued to the bitter end, and the service closed for good at the same time as the maritime service at 1200 GMT on 30th April 2000.

The station was, for many years, a major player in providing communication facilities for yacht races. Starting with the already-mentioned Golden Globe race in 1968, the station provided regular communications both to and from competing yachts well into the 1990s, when satellite communications became the standard. As the cost of installing HF radio equipment became more affordable on leisure vessels, it became clear that there would be a lucrative market for providing an HF radio communication service.

In 1976, the Post Office sponsored Clare Francis, skipper of the yacht *'ADC Accutrac'* in the 1977/78 Whitbread Round the World Yacht Race. She was the first female skipper of a competing yacht and was invited to the station for training and R/T familiarisation. The yacht completed the race in a very respectable fifth place. She is now probably better known as a successful novelist, writing many best-sellers.

Clare Francis at the Portishead R/T consoles with Ray Stevens and Neville Edwards.

Many famous yachtsmen became regular customers of the station: Chay Blyth, Robin Knox-Johnston, Pete Goss, Mike Golding, Lisa Clayton, Tracey Edwards, and many others all used the Portishead R/T services for both personal and position reporting calls.

British Telecom International (BTI) sponsored the yacht *'Hello World'* in the 1981 Observer-Europe Transatlantic Race, crewed by Eve Bonham and Diana Thomas-Ellam. Both crew members visited the station to familiarise themselves with the services and procedures, and BT was quick to cash in on the associated publicity with numerous articles and interviews. The vessel was apparently the smallest yacht ever fitted with radiotelephone and radio telex facilities.

'Hello World' publicity card, 1981.

Another race that Portishead Radio was heavily involved in was the 1985 Caribbean Race. The organisers (The Royal Ocean Racing Club) were quick to thank the station:

"I cannot thank you enough for this wonderful service which you and your team have provided. Naturally from the safety aspect alone, your service has been invaluable, but the interest created by the position reports has added a further dimension to the race for all of us here in the Club, not to mention the families and friends of the participating crews".

The 'Global Challenge' series of yacht races was set up in 1989 by Chay Blyth, and were held every 4 years, each with different sponsors. The first race was held in 1992/93 as was known as the 'British Steel Challenge'. Ten yachts took part and all used Portishead as their main communications centre. The station provided all competitors with propagation charts to assist in choosing the optimum radio frequency to contact the station, and representatives from the station briefed the crews prior to the start of the race. All yachts successfully completed the race, which was won by *'Nuclear Electric'*, skippered by Paul Chittenden.

Following the success of this event, the next race in 1996/97 was sponsored by BT, and became the 'BT Global Challenge'. No expense was spared as BT supplied equipment, value-added services, and satellite position and tracking equipment. Portishead was again heavily involved, despite the use of Inmarsat-C equipment as the primary communications method. This time, 14 yachts competed and again all successfully completed the race. The winner on this occasion was *'Group 4'*, skippered by Mike Golding.

Sadly, this was to be the last of these races which Portishead was involved in; the next event (again sponsored by BT) was held in 2000, by which time the station had closed and all communications were satellite-based.

Other high-profile races used Portishead's services; the 'Clipper Round The World Yacht Race', organised by Clipper Ventures was a popular race and well supported by Portishead. Clipper Ventures was the brainchild of Robin Knox-Johnston and used yachts of equal size and design, crewed by amateurs. These races started in 1996 and continue to this day.

Another famous round the world race was the Whitbread race, started in 1993 and which evolved into the Volvo Ocean Race in 2001. Again, Portishead was heavily involved in providing radiotelephone communications to and from the competing vessels until the station closed.

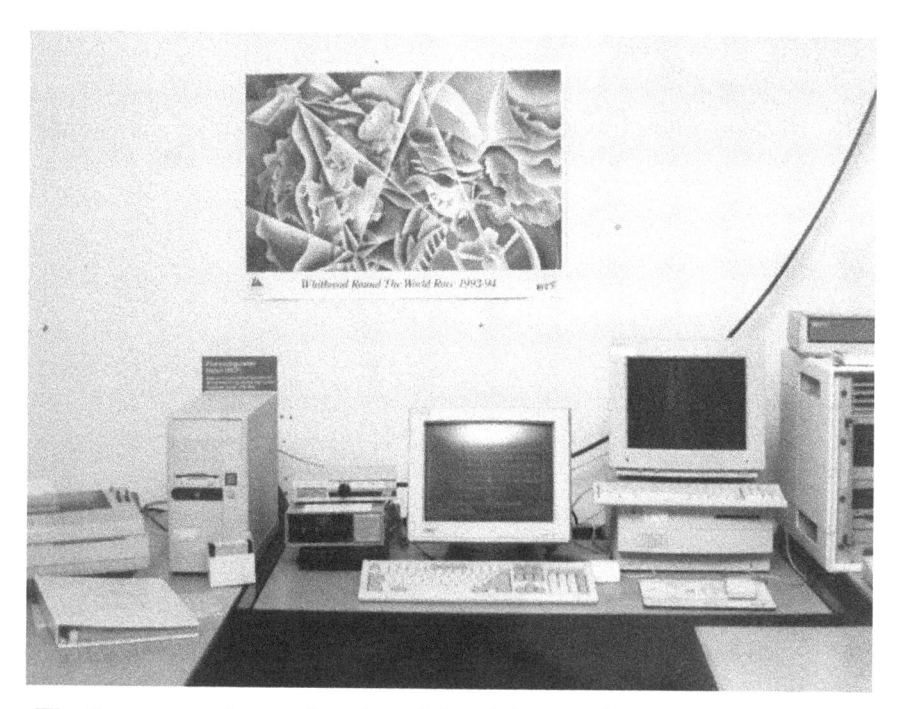

The Inmarsat-C console at Portishead for handling communications with the Whitbread yachts in 1993/94.

The Transatlantic single-handed race was another race that Portishead was heavily involved in. Again crew briefings and radio propagation charts were produced, and advice given on an individual basis when required. I recall visiting the competing yachts in Plymouth in 1996, with one of them not even fitted with maritime radio equipment. The organisers were rightly concerned about this but I was able to convince them that communication with Portishead was available by using the aeronautical radio channels. So for that one race, a competitor regularly reported on aero frequencies, and indeed successfully completed the voyage. That particular race was sponsored by 'Europe 1', a French radio broadcast station, and as such there were plenty of representatives from St. Lys (the French maritime radio station) who were vying for business with Portishead. Amusingly and thankfully, many of the British yachts refused their custom.

For all the above races, those yachts fitted with Inmarsat-C equipment could receive Phonetex messages which would be telephoned, faxed or telexed to Portishead for onward transmission to the desired yacht.

The leisure market was important for the station. Useful guides detailing HF communications for yachts were regularly produced, as well as laminated information sheets showing all the UK coast radio stations and their VHF Channels.

BT Radio Station – Portishead Radio
HF Radiotelephony Service

Portishead Radio monitors the following main assigned channels:

Callsign	ITU channel	Ship receive	Ship transmit
GKT20	410	4384	4092
GKU46	816	8764	8240
GKV54	1224	13146	12299
GKT62	1602	17245	16363
GKT18	1801	19755	18780
GKT76	2206	22711	22015

Please make your initial call to Portishead Radio on one of these channels.

Traffic lists are broadcast on these channels every hour on the hour.

Portishead Radio also operates a number of secondary channels. These will be announced on the main channel when they are available.

- A 3-minute minimum charge applies to all calls
- Calls to the UK may be billed to the destination telephone number. Ask for a 'direct billing' call
- Cheap rate calls available at weekends if billed to your accounting authority (AAIC)
- Transfer-charge calls available to the USA and Canada
- Medical advice (medico) information available free of charge

BT *The 'key' to your communications needs*

The credit card-sized information cards sent to leisure market users.

Guides for yachtsmen were also produced, written by Portishead R/O Charles Mander with illustrations provided by Portishead's in-house cartoonist Joe McCabe, which proved extremely popular.

Maritime Radio Services for Yachts and other Small Craft

External Telecommunications

BT's guide to Maritime Radio Services for the leisure market.

Many yachtsmen visited the station and were made most welcome, and during the 1990s the station ran free one-day HF radio familiarisation courses for yachtsmen, which proved extremely popular. Each morning, a

Radio Officer would give information and advice for calling and working the station, and the afternoon session was spent 'sitting in' with the radiotelephone section, seeing the service in action. Indeed the station was the proud recipient of a Royal Yachting Association award in 1995 for services to the leisure market as a result of these courses.

Yachts continued to use the R/T service (and, in some instances, the radio telex service, which had become affordable to the leisure market) until the station closed. A popular and much-missed service to the leisure community. Yacht owners were encouraged to visit the station, and many took up the invitation.

Sadly, the days of the station were numbered, and when the station finally closed in 2000 many yachtsmen mourned its passing. A friendly voice on many a dark night had gone forever.

CHAPTER NINE

LIFE AT PORTISHEAD RADIO

With over a thousand people who have worked within these hallowed walls, there have naturally been more than one or two 'characters', about whom many a tale has been told. Some stories have been expanded somewhat, whilst others have attained legendary status.

There have even been stories which one cannot (for obvious reasons) divulge outside the station, but some of the more amusing and unusual tales can now be related.

Morse code - a simple enough method of communication, one would have thought, especially for a 'professional' Radio Officer. Alas, there are (and always have been) some R/Os whose standard of Morse have left a lot to be desired. There was nothing more frustrating than trying to communicate with an operator whose command of the Morse code was somewhat sketchy, and sometimes this frustration led to boiling point.

There are many examples of Portishead Radio operators who, having sent hundreds of words to a ship only to be greeted by the reply 'please repeat all', have muttered a few expletives and thumped the desk in sheer exasperation. There are also examples of operators who have taken the law into their own hands by hurling their typewriters out of the window (luckily open at the time) or venting their fury on the unsuspecting station cat. However, the majority of the operators managed to contain their wrath surprisingly well, and the more 'laid back' operator would simply raise an eyebrow, take a deep breath, curse internally, and get on with the job, no doubt silently questioning the parentage of the Radio Officer at the other end....

Radio Officers have their own language - and different ways of sending the Morse code. It is quite possible to send 'sarcastic' Morse (by extending the dashes), or to send 'impatient' Morse (clipping the characters). Each operator, especially in the days before electronic or automatic Morse keys, had his or her own sending characteristic, and it was quite possible to identify a ship simply by the Morse code sent by the ship's Radio Officer. Operators at Portishead could use either a traditional 'up and down' key or an electronic 'paddle' key. Some actually preferred to use their own keys, such as the semi-automatic 'Vibroplex' or a paddle keyer of their own, but again, each operator had his or her own characteristic, and it became commonplace for a ship's Radio Officer to 'know' the Portishead operator simply by the style of Morse being sent.

Apart from the internationally agreed 'Q' codes, which were in everyday use in the W/T service, Radio Officers tended to use their own abbreviations to communicate with each other. For example' 'TU' meant 'Thank You', '73' meant 'Best Wishes', and 'OM' meant 'Old Man'. Radio Officers have always used 'Old Man' as a courtesy - much less informal than 'Sir'. Even female Radio Officers were referred to as 'Old Man' in the early days!). A lot of abbreviations were simply shorthand ways of sending commonly-used pleasantries, and many could easily be decoded.

It was the practice, years ago, to log the ship's name and call sign if the Radio Officer on board was guilty of poor operating procedure or badly-formed Morse. This was to ensure that if a mistake was made in a telegram due to either of the above, the ship's owner or agent could be advised and necessary action taken. This log was ceased some years ago, when a ship's Radio Officer, on leave from his ship, was being shown around the station and for some reason shown the 'bad Morse' book. There, clear as day, on the last page, was the name and call sign of this officer's last ship, which needless to say did not go down too well with the officer concerned. Subsequently, all radiotelegrams were taken 'as sent', and only in cases of extreme ineptitude were the ship's owners advised. Some of our more pragmatic Radio Officers referred to the W/T service as 'radio telepathy' in the final days of the service, due to the proliferation of poor Morse at sea.

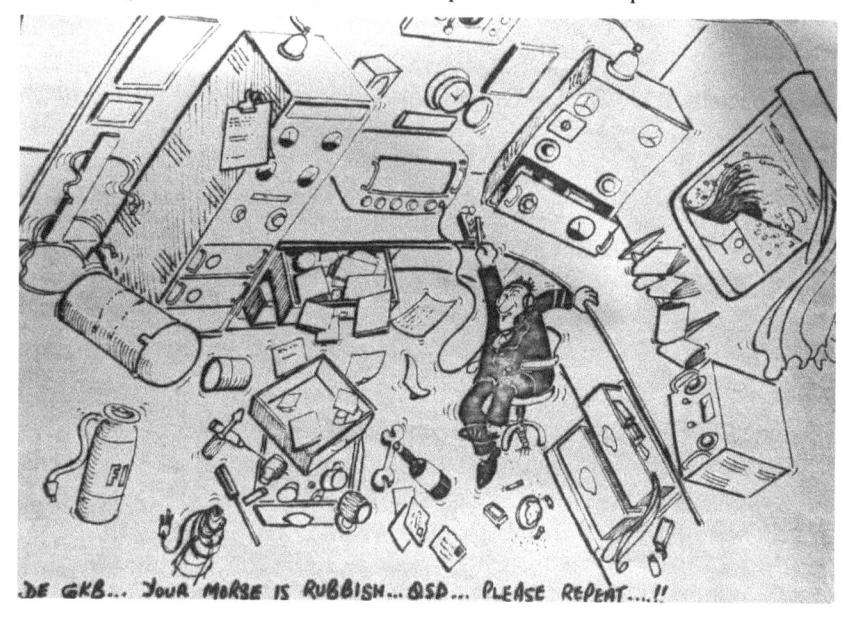

DE GKB... YOUR MORSE IS RUBBISH...QSD... PLEASE REPEAT....!!

Shipping companies have been aware of the fact that words were charged in 10-letter groups for some time, and many of them used this fact to their advantage by abbreviating and running together words. For instance, the phrase 'PLEASE ADVISE YOUR ESTIMATED TIME OF ARRIVAL AT LISBON' (10 words) would become PLSADV/ETA LISBON (2 words), or 'BUNKERS REMAINING ON BOARD' (4 words) would become 'BROB' (1 word). These abbreviations, in fact, are now commonplace amongst the shipping world, and indeed some companies use their own code words for the more common phrases. A couple of enterprising companies even tried abbreviating phrases to the minimum number of characters physically possible, causing confusion at both ends.

There is, however, some danger in abbreviating telegrams in this manner. I well remember a telegram which began 'SHIP/IS/OLD AND/MUST PROCEED SLOWSPEED etc.' A telegram from the ship later that day queried the first group and asked to whom the ship had been sold! Obviously, the Radio Officer had taken the first word as 'SHIP/IS/SOLD' and did not query this with our operator. This episode took a couple of telegram exchanges to clarify, and all for the sake of one character!

Radio Officer Gil Elks recounts one particular exchange of traffic with one of the Union Castle ships;

"A colleague was handling the contact when the ship sent a message with the text containing just one word – IMPOSSIBLE.

The following exchange then ensued;

PORTISHEAD:	Please repeat text.
SHIP:	Impossible.
PORTISHEAD:	Why is it impossible?
SHIP:	The text is I M P O S S I B L E
PORTISHEAD:	Is the writing that bad you cannot read it?
SHIP:	Look the text is I M P O S S I B L E
PORTISHEAD:	I understand the text is impossible to transmit.
SHIP:	Yes the text is IMPOSSIBLE and so are you please QTA (cancel)".

Some telegrams have been unintentionally amusing, and one had to look twice at the text before sending it to the ship. One telegram (which came from Ireland) many years ago stated that 'CREW WILL NO LONGER BE PAID FORTNIGHTLY STOP INSTEAD PAYMENT WILL BE MADE EVERY TWO WEEKS STOP'. The reply from the ship was unfortunately not recorded.

One of the most-remembered episodes of the 1980s came from Texaco in London. Again, in order to cut costs, all ships in the Texaco fleet were instructed not to use the word 'REGARDS' at the end of their telegrams – simply sign off with the word 'MASTER'. After a long explanation as to why the word 'REGARDS' should be omitted, the telegram was signed 'REGARDS TEXACO'.

It was vital to ensure that what was sent by the shipboard Radio Officer was correctly received at Portishead, even if it appeared at first glance to be incorrect. Radio Officer Len Wilson, at sea between 1957 and 1961 recalls the importance of double-checking;

"I recall on one occasion preparing to send traffic to London via Hong Kong Radio when we were in Indonesian waters. Whilst spinning the knob to tune in Hong Kong I heard Portishead. Changing my tactics I made a quick call, got an immediate response, and sent the message direct.

The story then takes an amusing turn. There was trouble in Indonesia at the time and a ship had been bombed, so we were instructed to report our position daily to Head Office. As I was congratulating myself on the speedy despatch of one such report, sent in company code, I decided that I really ought to check the code for the 'Old Man' had been taking his 'medicine' fairly regularly. The result was the cause of a great deal of mirth and hilarity, for he had used the code for latitude South instead of North, putting us halfway up a mountain on one of the islands. To save his skin a correct version was sent immediately, and I was delegated the duty of encoding all subsequent company messages".

One nameless Radio Officer recalls;

"One episode which stays in my mind was when I sent a telegram to a ship saying 'Baby boy Friday. Both well - love Charlie'. The ship asked me to wait then came back saying the new Dad was delighted but wanted to know 'who the hell Charlie was?' I went back to the origin and found that the message had been phoned through to the local post office. The signature should have read 'Shirley'.

Amusing incidents were of course not confined to the key; the Radio Telephone service had its fair share of 'magic moments' too. One incident recalled by Mr. G.W. Griffin illustrates this;

"Delivering a large yacht to the Caribbean many years ago, I regularly kept in touch with home via Portishead Radio. On one occasion I was calling my wife from Mid-Atlantic and so as not to occupy the airwaves for too long, she promised my 4-year-old son that he could tell Daddy just one thing. Broadcast over thousands of square miles of ocean came a breathless little voice, 'Daddy I can wipe my own bottom now'!"

Then there was the British R/O on a British ship who called Rogaland Radio on RT and asked for the traffic list times. On being told that the foreign traffic lists were at such and such a time, he said: "Look here old man I'm not foreign, I'm British."

During the 1970s and early 1980s, staff were transported to and from Somerton by minibus to man the maritime R/T circuits. This could be an 'experience' as recalled by Ramsay Stuart:

"Regarding the Somerton bus – There were two drivers, Ron Westlake (aka 'Rapid Ron') being one, and the other Harry Brown. It was Harry who liked to vary the route a bit. The standard route was left at the mini roundabout on Church Street, across the railway, turn right at Watchfield, up through Woolavington, left on the Bath Road (A39) to the Pipers, right and up Walton Hill and Ivythorn Hill, right on the B3151 and right on the B3153 to Somerton.

On a June morning, the view from Ivythorn Hill was spectacular. There was usually a low hanging mist and Glastonbury seemed to be an island floating in the clouds.

On one occasion Harry turned right on the A39 at The Albion, down the hill, right on the A361 to Pedwell then left and across Kings Sedge Moor to Low Ham thence through Pedwell (where there were guinea fowl running about in the road – talk about rural – almost as rural as Nempnett Thrubwell) up to the B3153 then left and up to Somerton. The rhynes alongside the road across the moor, lined with ancient pollarded willows, are something to behold – a double-decker bus would disappear if it fell in. It did not take but a few minutes longer that way, but it was a welcome change of scenery.

Duty at Somerton attracted a daily subsistence allowance, which was enough to buy a decent lunch and two pints of bitter at the White Hart, or the Globe next door. Better fare was to be had at the Half Moon, which was Egon Ronay rated for the cold buffet, but although the grub was marvellous it did not also run to a couple of pints.

When on duty at Somerton we had the use of a vehicle to get down into town and get a meal. It was an estate car or station wagon or whatever you like to call it, but one day there was a new one, and when we drove it we were not familiar with the controls, therefore it was not unusual to see the vehicle going down the road with hazard lights flashing, headlights on, and windscreen wipers operating.

The landlord of the White Hart was a miserable little squirrel, and, although most of us liked the place, a number of incidents turned us off. I may have got this wrong, but Phil Lewis, for I think it was he, or perhaps it was Nigel Le Gresley, asked for a cup of tea, and was rewarded with a heap of verbal abuse. Most of us then transferred our custom to the Globe next door to the White Hart. The Globe was something else. It was manned at lunchtime by three ladies who were, I believe, the wives of RN personnel at Yeovilton who were the licencees.

The Globe was brilliant – you could get a 'beefburger grill' (which was a decent mixed grill but with a beefburger instead of a steak) and a couple of pints on your allowance. If you didn't eat it all, the cook would come out of the kitchen and demand to know what was wrong with it! There was an old gent who went in every day at lunchtime and had a really huge bowl of soup and a loaf of bread – I think the ladies must have quietly subsidised him. The two ladies behind the bar were overheard discussing the cook, a rather large lady. "Do you know?" one said to the other, "She came in a wrap-round dress which wouldn't have wrapped around a bloody pencil."

If a stranger came into the Globe and asked for a sandwich he would then be asked: "Have you seen one of our sandwiches?" If no, then the ladies would demonstrate – you would have to have had a huge bite to get your face around it.

One day, I think it was a Saturday, while the roof of the Globe was stripped for re-slating, the roof collapsed and fell into the bar. It was out of hours at the time and the ladies and their kids were in the garden at the back, so no-one was hurt. It seems that some bodger a couple of hundred years ago had made a crappy scarf joint in one of the beams. Anyway, the beams were replaced with massive elm beams, thanks to Dutch elm disease which had released a huge amount of timber. There was nothing wrong with the timber – it was only the trees that died."

Sometimes the minibus would arrive back at the Highbridge site before 2300, causing one particular officious overseer to request the staff to work a few ships on W/T before the night shift arrived. These requests were 'politely' declined. However, one particular R/O, Phil Murray, was the exception to the rule, and he regularly volunteered to take a few ships before the end of the shift. Phil was a very religious man and even refused to go home should he win the 'draw' on a night shift due to his opposition to gambling.

Frank Ryan recalls a couple of Somerton-related episodes:

"When Somerton first started operating the secondary channels on R/T, only 3 went across for the day duty. I remember with Stuart Lund and another staff member, we had an "extended" liquid lunch in one of the pubs - we got back late to find Station Manager Don Mulholland and some Japanese gents awaiting our return. Don was not best pleased, but did not issue a "skin", or make any further mention of this.

When an Overseer was instructed to attend for the day at Somerton, Russ Taylor used to turn up with his wife at about 9 am, hang about until 9.30, then casually mention he was going into town - not to be seen again! Station Manager Arthur Hamblin kept on ringing up one day every half hour to try and get Russ. In the end, we stopped making excuses and said he had disappeared. Never did find out what happened."

Whilst on the subject of driving, access to and from the Highbridge site (especially by car) wasn't the easiest. There was a narrow lane (Worston Road/Worston Lane) which was the only direct route from Burnham to the station, but this wasn't the safest of routes for driving – and in fact was made one-way only (Highbridge to Burnham) during the 1970s. This meant that staff driving from Burnham had to drive to the A38 at Highbridge and take a left turn at the railway bridge to follow a narrow road to the station. Staff from the northern approach could negotiate 'Pople's Bow' which was another narrow lane from the A38 which had a bridge over the railway followed by a tight bend. For many staff, cycling to work was the preferred option, although the local constabulary were quick off the mark to check that one's bicycle lights were in order, and that cycling down the one-way lane was in the correct direction.

Before the welfare club opened it was customary for those who liked to imbibe to share a car and travel up Worston Lane to spend a grace relief in the 'Lighthouse Inn'. The more determined could sink three pints in the fifteen minutes available (which included travelling time). One icy night, a certain Ford Cortina set out and failed to negotiate the first bend, spinning on black ice and finished up at a 45-degree angle in the ditch. Unhurt and undeterred, the occupants ran back to the car park, took a reserve vehicle and still managed to get the three pints in. The Cortina started first time when the tow truck dragged it out the following day.

Once the new housing estate near to the station was completed in the early 1980s, access was made easier with the building of Pepperall Road which cut off the A38 access route. There was also an access road constructed to the outskirts of Burnham itself, but sadly this was not completed until after the station closed in 2000.

Back to the station itself, don't forget that many radiotelephone channels were shared with other Coast Radio stations; those of us who monitored Portishead's two main 12 MHz channels will no doubt recall problems with Kaliningrad Radio on GKT51 and Monaco Radio on GKV54. Obviously, this caused a great deal of frustration for all concerned, an example of which (thanks to Chief Officer M.J. Coventry – on board the NERC vessel *'John Murray'*) can be related below;

"On one trip we were cruising in the Western Approaches and used Portishead as our closest coast station. At four o'clock one morning, on coming on watch, I was told by the second mate that a Greek ship had spent his entire watch unavailingly calling Lisbon Radio. The general antagonism to this constant repetition of the call was almost palpable, the more so as the Greek had been politely told by another ship that Lisbon Radio did not operate at night. Shortly after I came on watch, and several more abortive calls to Lisbon Radio later, an unidentified voice, which I recognised as Portishead, conducted a short conversation, thus;

"Ship calling Lisbon Radio"

"Oh, Lisbon Radio, this is Greek ship…" (with great joy)

"Greek ship….. why don't you shut up and f*****g well go to bed?"

The silence was deafening and the general relief was unmistakable. Unfortunately, this broadside was not fatal, and the calls to Lisbon Radio continued after a short pause".

Another problem with interference on shared radiotelephone channels occurred with Bern Radio/HEB from Switzerland. On one occasion the station manager was called by a representative from Bern, asking us to vacate the radio channel. The manager's legendary reply was along the lines of "why don't you go back to making cuckoo clocks and leave the radio communication to the experts"….

Whilst speaking on the telephone, it is extremely easy to forget that what you say will be transmitted all around the world, and can be listened to by other ships and 'casual listeners' at any time. An incident recalled by Mrs. F Brown of Edinburgh illustrates this problem;

"I remember how delighted I used to be when I heard the Portishead operator linking me to my husband, an officer at that time in the Merchant Navy. Not long after we were married, he phoned via Portishead and in my innocence, I asked: "is anyone listening?" "No, no one" was the convincing reply, and I proceeded to make the ship's R/O's night (and every other R/O for miles) as I unburdened my loneliness and love – I still blush when I think of it!"

Having partners away at sea for long periods at a time can naturally put a strain on relationships, and I can vividly recall being in charge of two R/T circuits at the same time; on the top receiver was a call between a husband and wife sorting out divorce proceedings, and on the bottom receiver was a call between an engaged couple sorting out marriage arrangements.

One new (non-seagoing) R/O had just completed his R/T training, and had just been called by the *'QE2/GBTT'*. Upon asking the vessel's position, the *'QE2'* replied 'just off Newfoundland'. A long silence from Portishead was only ended when the R/O on the *'QE2'* broke in 'you don't know where that is, do you?' – One red-faced young R/O took a lot of stick for weeks afterwards.

John Banham, ex-Master of the tug *'Redoubtable'* remembers one memorable incident;

"On one of my calls home, the operator digressed from the usual formal announcement "you are through, Redoubtable" – on one occasion I was told, "your good lady is on the line and so is your dog". I thanked him and spoke to my wife. At the end of the call, I thanked the operator for the reference to my dog, as this was a pleasant departure from the normal routine and so was most enjoyable.

He then gave me the reason for his pleasant comments by saying it reminded him of when he was at Humber Radio and connecting a homecoming trawler skipper with his wife. When the lady lifted her phone there was heard the loud barking of what seemed to be an Alsatian dog. It took some time before the dog could be cleared away before the skipper and his wife could speak together. Immediately she broke into a tirade of a list of things which had gone wrong at home, leaving him in no doubt of her frustration and annoyance. After this had gone on for a few minutes, the skipper interrupted by saying he was fed up listening to her, and would she please put the dog back on".

There was always some degree of rivalry between the Naval and Merchant Radio Officers, the RN operators being occasionally unsure about civilian operating procedures. One example of this is described by David Wear, a BP Tanker Radio Officer during the late 1970s;

"I was waiting my turn for an R/T call, so only heard the Portishead side of the conversation which went as follows:

- OK Warship 'X', what is your position, please?
- What - is - your - position?
- WHERE ARE YOU?
- No, no, that's no good, the Indian Ocean is a big place!
- All you merchant ships stop laughing!!"

A similar episode was handled by myself on one occasion during the mid-1980s, when I heard a warship calling on 12 MHz during the evening shift. The conversation between us (as far as I can recall) went as follows;

PORTISHEAD:	Your signal is not so good, can you give me your approximate position, please?
WARSHIP:	South of Biscay.
PORTISHEAD:	OK that's fine, can you go to the main channel on 4 MHz please?

WARSHIP: Going up….

There then followed a period of around five minutes whilst I retuned the transmitter and receiver to the 4 MHz main channel, but nothing was heard from the warship at all. This was surprising as I had previously worked a vessel off the Spanish coast on 4 MHz without any problems.

After a further few minutes of futile calling, I advised the warship to revert to 12 MHz. Thankfully, he heard me and we managed to re-establish contact. As I still could not believe that the 4 MHz band could not provide a suitable circuit, I asked the warship for a more precise position – to which he replied 'South Georgia' (in the South Atlantic). True, this was certainly 'South of Biscay', but not very helpful in setting up a quality radiotelephone link.

GKA R/O Ainsley Dalrymple was on an R/T point where there were several ships calling over each other for a turn making it impossible to distinguish any call signs. Ainsley announced in a very authoritarian voice;

"All ships stop calling. All ships stop calling" (silence ensued)

"Except the one with the blue funnel"

We'll never know if the ship R/Os checked the funnel or if the first ship to re-call really had one but it sorted the problem.

A regular warship on R/T was *'HMS Hermione'* who we referred to as the *'Hermi One'* – a name which some of the R/Os on board the vessel used to take great exception to. And why we continued to use it.

Even Radiotelex was not exempt from a few strange moments. In the mid-1980s, a new automatic system (BART – Burnham Automatic Radiotelex) was introduced, at that time the most advanced system of its type in the world. All a ship had to do was to call up Portishead on the appropriate channel, key in the telex number required (prefixed by the abbreviation DIRTLX for direct connection or TLX for store-and-forward connection), and send the message.

To end the message, the code 'NNNN' had to be used, which would trigger the telex answerbacks automatically; the BART system would then provide a date and time group with the duration of the call. If the ship wished to contact an operator for assistance, the code OPR+ had to be entered and to clear the circuit with Portishead, the signal 'KKKK' triggered the clearing sequence.

R/O Steve Fell remembers one particular vessel having extreme difficulties with this (in theory) simple system;

"I recall one vessel coming up when BART had just come in. He typed in OPR+ to contact the operator and typing;

OM I AM HAVING TROUBLE, AFTER I TYPE KKKK

OM I AM HAVING TROUBLE, WHEN I SAY KKKK

OM I AM HAVING TROUBLE – I KEEP DROPPING OUT BEFORE MY MESSAGE IS SENT IT SEEMS TO HAPPEN AFTER I SEND KKKK

OM ETC… ETC…"

(OM is the standard courtesy abbreviation, meaning 'Old Man')

As in every place of employment throughout the world, there are always a few 'practical jokes' taking place – the vast majority (at least it is hoped) in good humour. Even in the early 1950s, when duty overseers ruled with an iron fist, collars and ties were compulsory, and Morse code transmissions were monitored for correct character formation, the Radio Officers' sense of humour was still to the fore. Maurice Broyd recalls one particular episode from 1953;

"A quiet Sunday afternoon, a group standing around in the old central circulation area. One R/O (whose name I'm not sure of, possibly Shearer), walking up and down the Land Line room. Bob Johnson, ex-Burnham and North Foreland, said: "let's have a laugh". He made out a message on an incoming 'green' (from-ship telegram form), in convincing-looking Chinese characters, date-stamped it and put it on the landline belt. Watching discreetly we saw this landline guy pick up this green, have a furtive glance around and quickly drop it back on the belt. After a couple of minutes, Bob (also landline duty) wandered in. "Anything doing?" he asks. "No" replies Shearer. "Oh, there's one here," says Bob, taking the Chinese green out of the end tray. In those days, the printers were Telegraph Manual Switching (TMS). You called Bristol Telegraph Office on the keyboard, gave them a destination code and they put you through. Bob strolled nonchalantly over to a printer, sat down, but with the machine in local mode appeared to call Bristol for a connection. He then rattled off on the keyboard for about the right time, made up a couple of start and finish answerback codes, stuck them on the green, signed it off, put it in the tray and wandered back into circulation. If the expression 'eyes out on stalks' ever range true, that was it! I don't believe he ever figured out how it was done and we had to leave circulation to collapse!"

Another occasion which he remembers towards the end of the 1960s occurred on W/T; "I was on 8 MHz at the time – I asked the Search Point over the intercom for another ship and was given GBTT, QTC250 (250 telegrams) or thereabouts. In those days one had to watch the light on the key. When it stopped flashing you went out to the ship. I told the *'QE2'* to go ahead but unbeknown to me, Geoff Perry (later of Post Office Maritime HQ), had grabbed the key and told the *'QE2'* to send the lot in one go!

Fortunately, he decided to ask me if I was getting them after about the first 30. What with QRM (interference) I hadn't of course"

Jack Robertson (1968-1972) remembers one strange incident whilst on circulation duty.

"After joining the staff of the 'sleeping giant' and just after my stint in the school learning to touch-type etc., I was sitting alone one afternoon at 'circulation' when I found among the traffic from ships worked from one of the wings, a TR form which had something like 113/2 on it – but no TR. Not knowing what this was supposed to mean I put it to one side meaning to ask someone more experienced than I was about it later. After a while, a second one turned up bearing the figures 138/3 or similar. To cut the story short, I eventually accumulated about half a dozen of these TR forms all with similar 'messages'.

When I was later relieved by a more senior R/O, whom I asked about these cryptic 'messages', he answered something like "Oh, that's old Charlie on point 26; he's listening to the cricket on BBC long wave and those are the latest scores in case you're interested!"

A couple of practical jokes were played on some of the younger and inexperienced staff, especially those who had recently received their R/T training. When R/T bookings were made in the to-ship direction, the appropriate call sign was used to include the ship in the next traffic list. If a call sign could not be located, the ship would be called by the name provided by the caller. This resulted in a couple of dubious names being called in a few R/T lists such as the *'Esso Blue'* or the *'Hoof Hearted'*, much to the amusement of the listening ships. However, one particular yacht was blessed with the name *'Shy Talk'* and was indeed the correct registered name. A personal favourite was the USA-registered yacht called the *'Wherethefogarewe'* (as in 'where the fog are we') which was great fun to quote on R/T.

On W/T, some names were nice and simple, such as the *'Iz'* or the *'Po'* – however, one memorable Greek vessel had the name of *'Agios Nikolaos Thalassoropos'* which was regularly shortened to *'A.N.T'*. Even call signs became memorable. We have already mentioned the *'Miranda/GULL'*, but there was the *'Sophie/SLUT'* and the *'Dart Atlantic/GOOF'*.

Some of the more 'old school' staff were sticklers for the rules. Ian Malcolm recalls an incident from the early 1950s;

"Broadcasts were sent at 20 wpm which I consider was exactly right as, after all, that was the speed necessary to qualify for the 2nd Class PMG. One day, however, Eric Macpherson, a close friend of mine, took it upon himself to transmit the broadcast at 25 wpm. He began with a statement informing the R/Os that the broadcast would be sent at 25 wpm and asked for comments! The first I heard of this was when, Archie Madeley, one of the older overseers, had a brief word with me as he headed for the door and his lunch. "He's just sent the bloody broadcast at 25 words a minute! I'm getting the hell out of here," blurted the agitated man. That exercise was never repeated although, strangely enough, I don't think Eric was reprimanded for such a flagrant breach of civil service discipline. One or two complimentary comments did come in, but, of course, there are always some people who have to show off!"

Another story recollected by Ian recalls;

"The hardest work I did at Portishead was when we operated a 'work to rule' for more money. When we were operating normally, we didn't operate according to the rule book i.e. Ship's call sign 3 times de GKL 3 times, etc. It was just as I've stated above; GNCS GKL sent only once with even the 'de' omitted. The big passenger liners were on the seas in these days and ships such as the *'QE'* (GBSS) and the *'QM'* (GBTT) had scores of telegrams to send and no time to waste. During the 'work to rule', it was GBSS 3 times de GKL 3 times, etc. and I could sense the amazement/disbelief of the guy at the other end when I worked my first passenger ship under these conditions. I could picture his face and what he said to a colleague in the radio room about this 'new boy/twit' at Portishead until it dawned on them all that they were dealing with a reformed Portishead. Complaints began to pour in from the shipping companies - just imagine the *'Queen Mary'* showing up at Southampton without Cunard knowing exactly when she would berth - no pilot in readiness, no trains to convey passengers up to London, no, etc. etc… And it was the same with the outgoing traffic. The gummed strips of telegrams were spewing in from teleprinters and coiling up in the tall elliptical metal containers beside them. It was controlled chaos if that's not a contradiction in terms! And it couldn't go on. A representative of the Inspector of Wireless Telegraphy arrived at the station and, somehow or other, the matter was resolved although I have no recollection of receiving more money. What I do remember is that during that 'work to rule' period of a week or so, I went home absolutely exhausted and, years later, when I taught Trade Unionism to my pupils, I was able to tell them from first-hand experience, that 'working to rule' should not be taken lightly! And while all this was going on, the sheep would be grazing in the field outside and 'Cokey', the pure white Persian cat which lived on the Station, might be sprawled across your log!"

Ex-Portishead Radio Officer Paul Durkin (a renowned practical joker) relates a couple of incidents – the names have been removed to protect the innocent;

"One of my colleagues was on 'phonos' (accepting telegrams by telephone) when I rang in on an outside line pretending to be Mr.

Goulandris, the Greek shipping magnate, with a message to one of his supertankers – all in Greek and each word 15 characters long. After many efforts to take down the message on a typewriter (I asked for continual deletions and additions to the text which meant a new sheet each time), he finally asked if he had the correct message. I then asked him to read it back phonetically (at which he drew a sharp intake of breath but carried on nevertheless). I then asked him to insert a further paragraph adding, in my own accent "Thanks a lot, that's fine"! I could hear the explosion as I ran from the telex room to the car park!

On another occasion, I was on nights and was scheduled to spend a couple of hours in landline. I wondered why the lights had not been switched on in that Wing, but soon understood. The night overseer was knocking out some zzzzs in the first aid room as I breezed into landline – only to be confronted by a large billy goat tethered on a long rope which charged the length of the room towards me! I never found out who did it…"

From the very early days, the operating staff at Portishead was exclusively male; it wasn't until the 1970s that female radio operating staff were encouraged to join the service. This proved to be a traumatic time for the older hands at Portishead, as one R/O (who wishes to remain anonymous) remarks;

"We had all served in the British Merchant Navy where cargo ship crews were exclusively male. Exceptionally, the Master's wife might accompany him. Portishead Radio continued the tradition, will all clerical and administrative work at the station being done by male R/Os and ex-R/Os. There was no canteen and no tea ladies. A tea service to the operating point was provided by the Handymen, who did the cleaning and other necessary jobs. The station engineers were of course men. It was like being in an exclusive male club.

And now it was threatened! We braced ourselves for a flood of young ladies who would change our way of life. In the event, the number of young ladies who joined us were few, and they adapted to our system rather than changed it. I did notice, however, that if I was doing a Radio Telephone duty and one of the girls was on an adjacent channel, she always managed to attract a queue of Greek ships waiting to make calls, even with my channel clear! One of the young ladies was asked if she could cook; "No, but I can fix a radio" she replied! Our enquirer immediately cooled off, daunted no doubt by a vision of domestic bliss, all radios working but no dinners!"

One famous incident (which for obvious reasons we will have to maintain the anonymity of the young lady concerned), occurred in the early 1980s. One British vessel, whose W/T transmitter had developed a fault, came up on R/T to dictate a radiotelegram. In those days, these were taken down on a typewriter on 'green' radiotelegram forms. Halfway through the message, the typewriter keys jammed, thereby making transcription impossible. One could imagine the look on the shipboard R/O's face when the reply came; "Sorry Old Man, my typewriter jammed – I'll have to finish you off by hand!"

Keith Watkins recalls another embarrassing moment:

"One evening, I was on RT duty, guarding the main 8 MHz RT channel. It would have been around about 2000 so 8 MHz propagation would have covered quite a good area.

I built up a QRY list of about 7 or 8 ships, then proceeded to work them off. As luck would have it, QRY1 was a warship so went on for about 45 minutes. After signing off with him, I began to work my way through the QRY list but, inevitably I suppose, the remaining six or 7 ships had all gone elsewhere during the long QSO with the warship.

When no ship answered my calls, I turned around to the guy next to me and said something like "after that long one, they've all f****d off and gone somewhere else".

At that point, my neighbour had a big grin on his face, leant across to my operating console and lifted the microphone switch to the 'off' position. My ill-judged comment would have gone halfway around the world."

Something very similar happened back in the 1950s. Back then the station provided an MF service with its GRL transmitter used mainly on telegraphy and only rarely on R/T. Despite this it was tested daily by the maintenance man, the usual practice being adopted i.e. huffing and puffing into the handset and noting the response on the aerial ammeter. One day, the maintenance man did this and got no reaction. He huffed and puffed some more, before turning to a colleague in a loud voice and declaring "This f****** thing's not working". Unfortunately, it was only the meter that had failed - his beautifully modulated voice rang out over the airwaves. Breaksea Light Vessel replied, "Oh yes it is!"

Changes in frequency on MF were performed by staff at the Portishead site, and communication was via an order wire and service Morse 'PEY' followed by the working frequency. Portishead then changed frequency. This key was, however, sited alongside the W/T broadcast key. On more than one occasion Portishead phoned to say that they, and not the listening ships, had received the latest traffic list.

Apart from Radio Officers, the Post Office (and subsequently BT) employed other ancillary staff such as Handymen and Cleaners, many of whom were unique characters in their own right. Such a person was Jack Lovibond (senior), who did not suffer fools gladly. Former Portishead R/O Phil Mitchell remembers;

"Jack was heard to welcome a station visitor to the old station front door, who asked to speak to the "Boss" – with – "When you'm passed me they'm all bosses".

Another incident Jack was involved in was also recalled;

"Jack had a club foot and used to come struggling up the stairs with a tray which held a dozen cups of tea. A heavy load. There was a tendency for the Operators to get up and crowd around him to get a cup. On one occasion he lost his temper, shouting 'Get back you buggers'. This had no effect so he hurled the trayful of cups, full of tea, into the centre of the room".

Another handyman remembered with affection was Sam Evans. Phil recalls one particular episode;

"Sam habitually raffled a Christmas turkey to the staff, which was usually supplemented by R/Os on detached duty from the quieter coast stations – Burnham used to be very – very hectic at Christmas! Over a period of years, the raffle had been won by the detached duty bods, not the Burnham staff. R/O Mac-something or other won the raffle yet again. Sam led his turkey on a lead through 'A' wing, proffered the lead to the R/O at his operating position and indicated that he should 'tow the bloody thing back to Stonehaven".

Sam also features in other escapades, some of which are related below;

"One of the engineers had a new house in St. Mary's Road and was, quite naturally seeking cuttings for his new garden. Sam went into the workshop asking if a certain engineer was on duty (knowing full well that he was not). Sam said 'that's a pity, I've got some plants for him and they won't last unless planted soon'. His target, the new householder, said he would like them and asked how they should be planted. He was told 'about a foot apart along the back fence'. This he did and soon had the straightest row of nettles in the whole of Somerset.

When Sam was on the early shift he used to cook his breakfast around 7.30 am. One of the R/Os on Accounts Duty used to come in early as it was 'job and finish'. He would have a snack with Sam. One morning there was a good smell coming from the oven. Sam said 'ever tasted grouse?' On being told 'no', he tore the leg off and said 'try it'. A bite was taken with the comment 'not bad'. It was a blackbird that Sam had found on the lawn. He had carefully plucked and dressed it for the oven.

In the early days, with the exception of the Officer-in-Charge and his assistant, the mode of transport was the bicycle, which in 1949, were not easy to come by. Sam was the main supplier. Where they came from remains a mystery. One R/O had a lady's style bike. New tyres were unaffordable so he used to cycle to and fro with tyres stuffed with newspaper".

Fred 'Bruv' Mitchell, one of the doormen, used to recount many tall tales of occasions when he had (apparently) driven the England football team coach, sparred with Henry Cooper, and taught Bobby Charlton all he knew. None of the staff believed him of course, but he used to enjoy telling us these stories. Upon arriving at work one Saturday evening at 1800, I was greeted by Fred. He told me he had just got back from the F.A. Cup final at Wembley, which unbeknown to him had gone into extra time! No wonder he was out of breath.

Another handyman, Charlie Cox, was an outgoing and friendly Bristolian and would do anything for anyone – unfortunately, his car maintenance skills weren't as reliable as he claimed. I once asked him to take a look at my car, an ageing Vauxhall Viva which was having trouble starting. He came round to my place and tinkered around with the engine for half an hour or so, got the engine going, and all seemed fine. I slipped him a fiver or so for his trouble and all was well......or so I thought.

Later that day (in readiness for a 5-11 or similar) I got into the car – and would it start? Not a hope. Cycled into work and managed to arrange a tow to a local garage the following day. The mechanic there did the usual sucking of teeth and said that someone has had a right go at this (or words to that effect). Spent the rest of the day in the garage having a full engine service...cost me a packet to get fixed. Cheers Charlie. I never asked him to repair my cars again.

During the early 1970s, the story goes that one day a small group of officials visited for a meeting with the Officer-in-Charge. The meeting took place in the Boss's office (next door to Writing Duty on the top floor of the old building). During the morning the boss must have suggested a refreshment break. At that time another long-serving civil servant (in the guise of one Fred Lovibond) was one of the station's handymen, whose domain was the kitchen where he and his mate(s) prepared and toured each main day/evening shift with the ubiquitous tea-trolley (an essential piece of civil service equipment) in those pre-restaurant days.

Anyway, the Boss, together with probably the assistant OC, maybe a senior overseer and the officials from London paused the meeting and Fred Lovibond, a local family man with rather blunt, rough-hewn character traits and who usually shot his opinions from the hip so to speak was summoned to attend the meeting. The Boss and several pairs of eyes turned to the door as Fred strode in - probably dispensing with a polite knock, and gruffly asked why he had been called (remember this was a time long before any PR or hospitality procedures existed). Fred was asked if he would provide cups of tea for the visitors and other members of the meeting. Apparently, there was a startled reaction from Fred who gathered himself together, looked around the room, locked eyes with the boss and said in his broad Somerset accent;

"Yer what! Yer all want me to serve yer tea? What d'ya think I am - yer f*****g butler!"

The London visitors were doubtless somewhat abashed but Fred probably came as no surprise to the others. I believe they did get their tea in the end (and Fred kept his job!)

Overseers were, of course, another breed entirely; in fact, when I organised the 75th Anniversary celebrations in 1995, a couple of old-timers asked if 'old so-and-so' was attending (the person being in question being a particularly unpopular overseer during the 1950s and 1960s). When being told "yes, he'll be here", I was politely advised that "in that case, I won't be". Grudges go back a long time in this job.

Of course, the majority of the duty overseers were firm but fair, and maintained the respect of the operators. I recall operating the 22 MHz Search Point in the old station, and not hearing any ships calling GKB. Suddenly there was a tap on the shoulder, and the duty overseer (Jack Todd) said three words to me – "East plus two" (one of the receiving aerial directions – two clicks past East). I selected the appropriate aerial to hear a couple of ships calling GKB from the Persian Gulf. With that, the overseer was gone, job done.

David Whitehead remembers another overseer: "R/O Bob Hartop was on a day duty on one occasion and the lunch break was 1300 to 1340. He went home for his lunch. About half-past one, the telephone rang on the overseer's desk. Idris Gibson answered. Bob's wife called to say "He's fallen asleep and I don't like to wake him"

Idris said, "That's ok, just leave him". She did. So we didn't see him till his next shift.

Idris was a kind-hearted soul who wouldn't do anybody a bad turn."

Well hello there...... didn't you used to be a supervisor at Burnham Radio...!

During one severe winter storm, all roads in the area were snowbound and travelling to work by car was not a viable (or safe) option. One stalwart R/O decided to walk to work. As he passed the house of one of his radio station colleagues (one of the overseers), a voice came out of the upstairs window; "Tell them I won't be in work today, I'm snowed in".

'Robbie' Roberts recalls an incident with another overseer who was not quite so diplomatic;

"I remember an occasion when Les Cuthbertson was pushing his bike, with a broken chain, down the lane to the station building. Passed by an overseer who said "Good Morning Les". When he arrived he found the 'friendly overseer' had booked him for being late".

Les was another renowned character, who once made the tea on a night shift with added Epsom salts as he was fed up with continually making the brew. This caused the staff to visit the W.C. an extra two or three times during the shift; needless to say, he was never asked to make the tea again.

One overseer (Russ Taylor) had a penchant for taking a few hours' sleep during the night shift. This incurred the wrath of many a night duty team on many occasions, so one night, just before he was due to wake up to bring up the higher frequency transmitters for the daylight shift, all the station lights were switched off and the whole shift hid outside the station building. One Radio Officer was seen to start his car in the car park to ensure the overseer concerned was woken up! One can only imagine the look of panic on his face when he entered the control room to find the whole building deserted.

Geoff Pople recalls another car park incident on a night shift:

"I remember one night shift about 0030, a certain overseer came hurtling down into the operating area from on high and we had to have a headcount – we were all a bit bemused. He demanded to know who had driven off a few minutes earlier. We were all there. Later it turned out someone had left their car in the car park and called back later to pick it up".

Jack Robertson remembers another 'amusing' overseer incident;

"There was a certain overseer, who shall remain nameless, who was always making rather silly mistakes. There were an unusually large number of Collective Call sign messages one Christmas and I think they were transmitted in a separate broadcast from the rest of the traffic. Needless to say, we were all rushed off our feet as it was only a few days before Christmas. After I'd finished a Collective Call sign broadcast (note the word broadcast as opposed to a two-way communication with a ship) one evening, I went into the control room to put the messages back in their pigeon-hole. The flustered Overseer intercepted me, took them from me, scrutinised them and then asked: "why haven't these been QSL'd (acknowledged) yet?"

A similar occurrence took place with R/O Bob Atkin. Once, Bob was pounding away sending a huge wad of SLTs to Interflora and was sat back having a smoke when the overseer came past with a flippant remark about the SLTs not getting there by themselves. Bob looked up and said, "I'm sending them so fast that I'm giving the girl at the other end time to catch up preparing the flowers and cards". The overseer responded with glee that "that's not what happens Bob, when they get there they........" before realising Bob was grinning whilst finishing his cigarette. The overseer strode away red-faced.

In the early days of the service, overseers ruled with an iron fist; those of us who moaned and groaned at the attitude of certain junior managers in the 1980s and 1990s were treated lightly. Glancing through the 'Record of Officers' books between 1910 and 1947 (which include all Coast Stations, not only Burnham), even the most 'minor' misdemeanour was treated severely, although some incidents do appear to warrant some sort of punishment. The following examples (from Burnham only – other coast stations fall outside the remit of this book) demonstrate this;

Wireless Officer C. W. Dawson was reprimanded and warned to use greater care in future for 'signalling office of destination as Plymouth instead of Cherbourg' (March 1931). Similarly. T.W. Norrish received a caution for 'entering the call sign of *'Narkunda'* instead of *'Naldena'* in a message of 17.7.32 and for failure to check sent particulars to ensure the message had been correctly transmitted'.

Other operators were also cautioned for 'falling asleep whilst on watch', and also for 'insubordinate attitudes' to management. Both informal and formal written warnings were issued, although it was very rare for any severe disciplinary action to be taken. One favourite target for the more 'jobsworth' overseers was those R/Os who exceeded their 15-minute 'Grace Reliefs'. On an evening shift, it was common for staff to visit the bar in the welfare club for a swift pint and a game of pool. If there was a social event going on there was always a queue at the bar, although the bar staff were requested to give priority to staff on their break (to the annoyance of those already waiting). On their return to the operations area, some staff were met by the duty overseer who had made a note of their departure and return times.

There was a fruit machine adjacent to the bar, and many staff members spent hours devising their own 'systems' on how to win, although the machine itself was changed regularly to prevent the 'systems' from being implemented. A quiz machine was also installed for a time before it became apparent that certain staff members made notes of all the correct answers, therefore ensuring regular payouts when next they played.

It was often said there was an expert on anything with the station staff. There were mechanics, electricians, gardeners, mathematicians and so on. In fact the W/T area was often subdivided into informal areas for radio amateurs, sports fans, stockbrokers and Open University students.

During the 1980s it was quite regular for groups to be shown around the station. On one occasion, 'A' Wing was extremely busy with each console fully manned, each R/O taking down messages on their typewriters, concentrating intently and gazing into space. One of the visitors was heard to ask the Station Manager (Don Mulholland) how many blind R/O's the station employed.

Night shifts at the station, certainly during the 1980s, tended to be quite hectic. To try to boost morale (and to keep the staff awake), a 'draw' took place at around 0400 to allow one member of staff to go home. All the names of the staff on duty were put into a hat and someone was selected to 'make the draw'. Sometimes, if the nights were very quiet, there would be further draws to allow further staff the chance to leave early. On one occasion, the draw took place a few times, and one disgruntled non-winning staff member took it upon himself to check the names in the hat, only to find his name wasn't even there. I understand that after a few well-chosen words he took himself home anyway.

Graham Powell has one lovely story about Russ Taylor being the overseer on a night shift:

"One night we told George Barnard (one of the R/Os) that he had won the draw. So off home he went. The problem was that we actually hadn't had a draw, so every time Russ appeared we had to pretend that he was still there but had gone somewhere. We managed to cover it up until the morning but I don't remember ever doing it again".

On Christmas Day, there was an hourly draw, commencing at 2300 at the start of the watch, with the last draw taking place at 0700. I am pleased to say that I did win the 2300 draw on one occasion but had to finish the ship I was working before I departed. For some reason, nobody would take the ship from me.

The 'draw' had been one of the perks on the night shift for many years, although it was at the discretion of the night overseer whether to hold one or not. Frank Kelley remembers how the draws were organised back in the 1940s and 1950s:

"Interesting that in the mid-fifties, names were drawn out of a hat to decide who went home early. In 1947, there was no set time and (depending on the overseer) one could be released when all the SLT's had been typed and enveloped, usually about 0530-0630. The draw was done by typing the nine initials into a perforator and gumming the tape into a continuous loop. This was run in the transmitter set to 240 words per minute. After a few revs, switch off and the initials under the peckers at rest was the lucky one. Certainly better than the hat system and probably the first random number selector at the time!"

Station Managers were not exempt from practical jokes. One particular SM had a penchant for wearing a distinctive over-sized bow-tie to work, only to find that one morning all the station uplighters, plants, and even waste bins had paper bow-ties attached to them. We never saw the bow-tie again.

There is also the story of an argument between the Assistant Station Manager and a staff member which ended up with the R/O hanging the ASM's jacket on the wall. The ASM was unfortunately wearing it at the time.

There were a few fractious episodes between some of the more excitable staff. Frank Ryan recalls the 'great fisticuffs' between Point 41 and Point 42 in the old station.

"After the famous "If you think you are man enough, come across and do it", Point 41 (nameless), swung a right-hander and promptly dropped Point 42. Phil Coates, the overseer, happened to walk through the door just at that moment, promptly turned on his heel and disappeared to the sanctuary of the Wendy House - probably the best and wisest thing he ever did."

He also recalls the incident where R/O Mike Ward had an altercation one evening with the visiting Skittle Team (which happened to be the Radio Station's 'other' team) in the bar. He issued the challenge "you and your best three men outside". At that moment, Mike's daughter, a Policewoman on the Royal Protection Duty at Gloucester, came in, put an arm-lock on Mike and frogmarched him out of the welfare block.

In April 1993, new flexible 'stalk' microphones were installed to replace the incongruous R/T headsets with monstrous mouthpieces currently being used. An official notice from the Station Manager went up on the board advising the staff of these new 'STORK' microphones, to which an anonymous R/O had appended the following;

> *"The staff are being asked to chews*
> *The type of mike they want to ewes*
> *To make it easier for us to torque*
> *They're going to give us all a stork!"*

The notice disappeared within hours.

The station building was in those reached by a narrow lane (Worston Lane) which these days is one-way only (from the radio station to Burnham-on-Sea). However, the rhyne which ran alongside the lane still remains, and on a dark evening can be a hazard to pedestrians and motorists alike. Many an R/O, suitably tanked up with alcoholic beverages from the station bar has ended up in the ditch or in the next field. One R/O remembers;

"The rhyne which ran from the station to the Worston Lane houses claimed many victims from the station. One such was Barry Jackson who used to walk to work in AA breeches (from his time with the AA) and a white roll-neck pullover. Staff who had finished at 2300 were cycling up the lane when they heard a plaintive voice saying "shine a light please". It was Barry, up to his waist in water, all efforts to drag himself out having failed. He had walked into the ditch in the fog and had been there since 2200. If the 2300 finishers had not passed he could have been there all night as very few people used the lane at that time of night.

Another night, three of us were going home to North Avenue, one had no rear-light so we thought that one of us each side of him along the Burnham Road would be OK. Not so – a police car bell ringing and blue lights flashing pulled up in front of us. The culprit expressed surprise saying it was on when he left work. The policeman, being a suspicious fellow, unscrewed the light – no bulb! He was fined five shillings, quite a lot of money at the time".

The station was lucky in that it had separate entry and exit barriers. The entrance barrier was operated by the use of an entry card which raised the barrier when inserted into the appropriate slot. If one had forgotten your card you had to alert the doorman who would raise the barrier by turning a key by the side of his chair, although this would normally be accompanied by a few well-chosen words. The exit barrier was raised by a push-button system, although those on bicycles could attempt a delicate manoeuvre to bypass it. However, on dark evenings or after a swift drink at the bar, many staff members forgot the barrier was there, and it was quite common to see the exit barrier lying on the floor in pieces the following day.

The staff notice board (as you will have already noticed) was always a target for practical jokes. Spoof notices, humorous additions to official notices and cartoons were all appended (normally during the night shift) with extra detail, much to the annoyance of local management.

One infamous notice (which will not be reproduced here) took the form of a school prospectus with certain R/Os (pupils) and managers (masters) being referred in not overly complimentary terms. It caused a great deal of amusement but it was also removed from the notice board very quickly – but not before many photocopies were made and distributed.

One R/O (who for obvious reasons shall remain nameless) once left a selection of 'films for the discerning gentleman' on an operating point after a shift. A notice was subsequently posted on the notice board asking the owner to collect such items, describing them as 'videos of a medical nature',

It was common practice for the telex rolls used at the station to be 'rewound' so that both sides of the paper could be used for reasons of economy. It was part of the duties of the Maintenance Officer (MO) to undertake this task. It wasn't long before a notice appeared on the board instructing the MOs to commence rewinding the toilet rolls in the staff lavatories for 'reasons of economy'.

Graham Haverson recalls one episode;

"A few of us used to leave our bags/cases containing headphones, Morse keys, etc. at the end of 'A' wing if we were coming back later that day. I did this one time after going off nights only to be woken by the police about 09:30 and taken post-haste back to the station in a dressing gown. Some naughty person had chalked 'BOMB' on the side of my little case and put it up on the pipes in the cross corridor opposite the overtime board.

The Army bomb disposal squad had also been called out and wanted to evacuate the station and remove the case to destroy it with a controlled explosion. When I got to the station to identify the case they told me not to touch it but fortunately, I was close enough to reach up and pull it down. Not very pleased police and army team but it solved the problem immediately! They were talking about charging me with a hoax but thankfully I had a lift home with Barrie Williams and he stated that he had seen me put the case down on our way out at 08:00! Phew!"

David Strickland confesses;

"It must have been around 1982, I had got into winemaking and had seen a recipe for 'rose petal wine'. I was living in Brunels Way then, only about 300 yards from the station.

Afraid the urge overtook and one night shift, I stepped out of the front door on my break and did a bit of gardening (some will remember the two sets of rose bushes at the entrance to the station) I managed to bag up most of them and drop them home, I don't know if anybody noticed the next morning, but I did get a couple of dozen bottles out of the petals".

The station looked out over a large field where the receiving aerials were located, and it was quite common practice for some staff members on a night shift to venture outside at dawn to pick some of the field mushrooms that were growing nearby. Many a splendid breakfast consisted of some fresh bread obtained from the local baker on the way home, accompanied by some recently-picked fresh mushrooms before going to bed for a well-earned rest.

Phonograms were not the most popular of jobs at the station. This involved either taking down telegrams for to-ship delivery from members of the public over the telephone or delivering from-ship telegrams to nominated shipping company representatives. However, there were a few memorable incidents. Bob Singleton remembers:

"On one occasion, Pat Dear and Chris Allen were doing this duty, and I was taking incoming telegrams. Pat was gamely trying to pass a Greek message to the wife of the company rep, as he wasn't in. The wife didn't speak English all that well, and was not very interested in taking stuff over the phone. After a lot of hassle, Pat suggested GKA could just forward it on telex and finished with something like "So you don't want the message now, OK?" As the woman replied "No - don't want message" Chris Allen leant into Pat's booth and ripped a piece of paper near the microphone then crumpled it up, causing much consternation, and a mild telling-off for Pat some time later, I believe."

George Davison apparently had a similar experience. Having tried to read a phonogram to a Greek wife, George could not get her to understand a particular word, George started re-reading the word letter by letter;

"No madam A repeat A, no no, A, Alpha, Alpha, that's A L P H A, Alpha.....no madam..."

Rich McMilan recalls another phonogram-related incident:

"I phoned a Southport number and explained it was Portishead Radio calling and I had a long message in Greek for (whatever telegraphic address was in front of me in the message). I emphasised the "long message" etc. bit hoping to be asked just to put it on telex as they often did.

He seemed a bit hesitant but said ok, I have a pen and paper, go ahead. I read out the long message, he wanted it read slowly and he got me to spell all of the words. He was English (not a Greek speaker) and it was a painful struggle to pass the message.

At the end, he said, "Thank you - but why are you delivering this message to me?" This confused me. It turned out that I had dialled the wrong number."

One personal incident which I must confess to is the total lack of 22 MHz radio telex traffic for a few days back in the mid-1980s. I was working a night shift on the Ocean Weather Service (OWS) position; not the most taxing of duties. At that time England were playing Australia at cricket in a test match so I thought it would be a good idea to listen to the commentary on the radio. Being at night I thought the 22 MHz receiver would not be used so I retuned it to BBC Radio 4 and continued to monitor the cricket on my headphones for the rest of the shift. As 0800 arrived, I packed up, handed over the OWS position to my relief and departed. When I returned to my next shift a couple of days later, I was made aware of an internal enquiry as to why there were no minutes recorded for the radio telex main channel (GKE7) on 22 MHz over the last 48 hours. Upon seeing this I went over to the receiver rack and noticed the 22 MHz receiver was indeed still on 198 kHz, the BBC Radio 4 frequency. I retuned the receiver to the correct frequency and, lo and behold, the traffic figures started to return. I thought it prudent to plead ignorance and the investigation drew to a close with inconclusive results. Until now.

The Radio Officer has always been unique amongst professions; it is felt that one had to be of a certain 'character' in order to cope with the type of work handled, together with the associated stress and trauma which occasionally cropped up. In dealing with the public, nobody (no matter who you were or what you did) could pull rank, and many customers of 'celebrity' status were brought down to earth by a caustic remark from a Portishead R/O.

Probably the most famous (or should it be infamous) incident in Portishead's history occurred during the early 1970s. Numerous versions of the same story abound, and even the identity of the R/O involved is still not certain. However, one Christmas day morning, a customer rang up wishing to speak to the Master of the '*QE2*'. At the time, all R/T calls had to be booked well in advance, in order to give as many ships as possible the opportunity to speak with their families; no bookings would be taken on Christmas Day itself. The Portishead R/O quite correctly refused to take the booking, only to be greeted by the reply "Do you know who I am?" When this question was replied to in a negative vein, the customer retorted "I am Sir Basil Smallpiece, Chairman of the Cunard Line".

Quick as a flash, the R/O replied: "I don't care if you are Sir Basil F*****g Brush, you still can't have the call".

This was not the end of the story; an official complaint was sent to the station manager, and whilst acknowledging the fact that the call could not be placed, a great dislike was taken in being compared to a 'furry rodent'.

A similar incident involved myself around 1984, when the singer Simon LeBon owned a racing yacht called '*Drum*'. I was on the R/T Coordinator

Duty, taking bookings and answering telephone queries, when a customer rang wanting to place a call to the '*Drum*'. At the time, the vessel was in the Indian Ocean, so I decided to route the call via Perth Radio, which I knew was in regular contact with the yacht. I took down the caller's telephone number, and asked for his name. "Simon LeBon" he replied, to which I replied (without realising who he was) "LeBon......strange name...how do you spell it?" The instant I had completed my reply I discovered to whom I was speaking, and bearing in mind I was running a mobile disco at the time (and had many of his records), the embarrassment was overwhelming. My leg was pulled incessantly for months afterwards.

Many celebrities used Portishead's services; Sir Richard Branson, on his Virgin Atlantic Challenger Balloon and Speedboat endeavours; Sir Ranulph Fiennes on his Transglobe expeditions; Sir Robin Knox-Johnston and Sir Chris Bonington on various expeditions, not to mention many more. The vast majority were a dream to work with, with one, songwriter Mike Batt, even writing and recording a song in tribute to the station, not surprisingly entitled 'Portishead Radio', on his 'Waves' Album. At the time he owned a yacht called the '*Braemar/MSZJ*' and was a regular customer on both radiotelephone and radio telex. The Portishead staff were pleased to receive a complimentary box of LPs as a mark of gratitude.

Radio Officers were also required on occasions, to act as an examiner for Radio Amateur Morse tests, at the (to us) excruciatingly slow speed of 12 words per minute. These were normally held on a Saturday afternoon in the training school, where up to four nervous wrecks would sit down and take perfect Morse from a tape reader/decoder at an exact 12 w.p.m. On a personal basis, I preferred to send the Morse by hand with a stopwatch in front of me, and I know that many of my colleagues did too.

It is amusing now to look back and remember the state these examinees would get into; one candidate would not get out of his car suffering from extreme nerves, and I recall another candidate taking the whole test without headphones – because no-one had told him to put them on! There was one memorable occasion when having successfully negotiated the test, the candidate asked if he could use the telephone; not to tell his nearest and dearest of his achievement, but to order over £1,000 worth of radio equipment to pick up on his way home.

I was threatened with violence on one occasion when I had cause to fail a candidate for too many receiving errors (they were allowed four), and he stormed out of the school without signing the appropriate forms, informing me that I was lucky not to get 'filled in'.

Another strange candidate had learnt the Morse code himself; unfortunately, the code he had learnt bore very little resemblance to the code devised by Samuel Morse, and an examination failure was inevitable. A similar incident occurred when a candidate had learnt his Morse code figures 'back to front' i.e. the dashes became dots and vice-versa. It was also not uncommon for candidates to break down in tears (either of grief or happiness) at the end of the examination, depending on the outcome.

Sometimes staff members were despatched to various locations around the UK to conduct Morse tests for radio clubs. I recall one occasion in Leeds when I was conducting a test and the building caretaker came storming in mid-way, informing me and the class to get out as he was locking up! Thankfully we managed to re-arrange that part of the test for the following evening.

On the whole, however, these were quite enjoyable diversions from the hustle and bustle of HF W/T working, and on most occasions, the candidates were given a fair hearing. For some reason, the Department of Trade decided to give the Morse code examination contract to the Radio Society of Great Britain at the end of the 1980s, which many people felt was a retrograde step.

There were quite a few radio amateurs amongst the staff. One, in particular, was very enthusiastic, and had suitable equipment installed in his car. On a break, he used to drive up to the seafront in Burnham-on-Sea to have a few quick contacts before returning to work. Another one managed to purchase part of one of the old aerial masts at the Highbridge site and erected it in his back garden, much to the annoyance of his neighbours. The same R/O also managed to find out that the quick-tune transmitters at Rugby would work on the amateur bands, so during quiet periods he used to fire them up and have a quick amateur contact before re-tuning to the assigned maritime frequency.

One night in the landline, Scheveningen Radio/PCH (who also had a good number of radio amateurs amongst their staff) rang up to enquire if there were any radio amateurs on that night. R/O Owen Kirby said we were all professionals and they put the phone down.

One of the great advantages of working at Portishead in the 1980s was the amount of overtime available. On many occasions, one could walk in at lunchtime, ask if there was anything going, and be told to 'get yer phones on' for the next few hours. The following week's overtime list would be displayed on the staff notice board each Wednesday morning, to be greeted with the cry of 'it's up!' and a subsequent scrummage to initial duties which may or may not be taken up by the nominated R/O.

As soon as the sheet was pinned up, many R/Os leapt from their point, diaries in hand, to prepare the next week's attendance. It was a 'gentlemen's agreement' that whoever initialled a person's listed overtime first would be given first refusal, although this did lead to some extreme methods. Some

R/Os would agree with certain others in advance to cover night duties etc., with cash changing hands on occasion! One poor incumbent, taken ill on station, suffered the indignity of being assisted home by one of his colleagues only after ensuring that his listed overtime had been re-allocated to his good samaritan! To an outsider, the whole system looked like the London Stock Exchange, with duty exchanges being arranged amongst a melee of cash-strapped Radio Officers.

Short notice overtime was given out daily by a member of Writing Duty, although it seemed strange to us 'new chaps' that the same people were getting allocated these duties regularly. It became clear that the 'old boys' network was alive and well on the station, with a certain amount of 'back-scratching' taking place! In fact, this situation became so intolerable that the local union and management became involved, in the end agreeing to dispense the overtime on a 'fair and equal' basis. For some strange reason, the same people still managed to get more than others. In the end, a listing of the amount of overtime performed by each member of staff was displayed on the notice board, so that if any overtime did come up, it was offered to the R/O at the bottom of the league table first.

Of course, as with any other large employer, there was a host of social and welfare activities available to staff. The station had its own welfare club and bar, and ran various sports and similar clubs. Indeed, during the 1970s and 1980s, the club ran successful teams in darts, skittles, pool, cricket, football, and rugby. In addition, the station ran one of the Post Office's drama clubs, winning the national drama competition in 1976. The drama club was organised by RADA-trained Radio Officer Charles Mander, a well-known author of one-act plays, some of which have been performed on television. Charles also wrote many of the Radio Station entries to the National competition, and although no further championships were won, various certificates of commendation were received. The club (known as the PRATS - Portishead Radio Amateur Theatrical Society) continued to flourish when Charles retired, and entered most of the National competitions until they ceased in the mid-1990s.

"5 To 11..... This evening ... 5 To 11..."

The welfare club itself was run by a committee of R/O volunteers under the auspices of the Burnham Radio Recreational And Welfare Committee (BRRAWC). The committee ran the bar, arranged functions and events, and was also responsible for the well being of staff. It also took on providing retirement gifts to staff after the usual collection of 'donations' depended on the popularity of the retiree concerned. One individual received around 10p in coins, a selection of washers and some foreign coins in his retirement collection, which prompted the committee to arrange a standard retirement gift for all staff.

The station also entered teams in the GPO/BT General Knowledge quizzes, and reached zonal finals on many occasions. The Radio Station team also did well in the local Quiz League in the late 1980s, although this folded when a team of professional 'quizzers' joined the league and won every week.

Football teams were entered into the local Sunday league, and a Cricket team was formed to play in the local evening league. The teams were also entered into the National Post Office/BT tournaments, and although they rarely progressed further than the early rounds, a few shock results came their way. One famous occasion occurred in the 1980s when the station Cricket team beat the Taunton Post Office team which included a number of county 2nd XI players and high-level Somerset league cricketers. Sadly the run did not continue, the team being dismissed for 21 against Exeter in the next round.

Kevin Cheeseman recalls a few episodes from the GKA football team:

"I well remember the incident at the Portishead Radio v Bridgwater Post Office football match in the SW PO competition many years ago. We had a stalwart team in those days including some nippy players and quick on the turn such as Frank Ryan (who martialled the defence well and played a good offside trap but unfortunately normally forgot to tell the rest of the defence), Ray Stevens, Vas Hira, Pat Dear and Rich McMilan (the last two who normally spent the second half of a match in the local hospital being stitched up). Graham Jones (another Radio Station R/O) was the referee that day.

Anyway as the match was about to start it was realised we were missing a linesman, so Jack Robertson (who had turned up with Chris Troy and a carrier bag full of beers) volunteered.

Jack took the flag from Graham and run to the other side of the pitch to take up his station for running the line.

Anyway, the game began and early in the match Bridgwater Post Office made a break from inside their own half and left the Portishead Radio defence for dead as one of the attackers from the other side bore down on goal with only our goalkeeper (Alan Linnett I think) to beat.

As the Bridgwater PO's striker moved closer in on goal the ref (Graham) saw Jack was frantically waving his flag so he immediately blew his whistle and stopped the game upon which the Bridgwater team was furious; how could their player be offside as he had broken from his own half? They were not happy, to say the least.

Well, Graham being the consummate referee agreed to check with Jack why he had flagged. Upon asking why he had been flagging, Jack looking rather sheepish, made his apologies as he was actually using his flag to try to attract Chris Troy's attention and bring him a can of beer.

The postscript to this match is that the Bridgwater Post Office team had a goalkeeper with one leg and he could only dive one way. Portishead Radio were given a penalty and to score it was a dead cert. The taker (who will remain nameless) only had to kick the ball to the side of the goal that the goalkeeper couldn't dive. I will say no more other than the penalty was saved Fortunately Portishead Radio did eventually run out 5-0 winners".

Another local match ended when the radio station team walked off after consistent fouling and abuse from the opposition. No referee was available for the match which ended in chaos.

"ITS CHRIS TROY.... HE JUMPED OFF THE GROUND WHEN WE SCORED, AND TWO QUARTS OF GUINNESS IN HIS TROUSER POCKETS SWUNG TOGETHER ON HIS GOOLIES!"

A highlight of each year from the early 1980s until the early 1990s was the annual football tournament, in which teams from some European Coast Radio stations regularly participated. Originally conceived in the 1970s for stations on the North Sea Coast, the tournament was expanded to include many other stations. At its peak, teams from Norddeich/DAN, Elbe-Weser/DAC, Kiel/DAO, Oostende/OST, Scheveningen/PCH, Portishead/GKA, Lyngby/OXZ, Rogaland/LGB, St. Lys/FFL and Bern/HEB participated. Spread over 3 days, the tournament was a good time for R/Os from each nation to get to know each other socially, and to visit each other's stations. Many friendships were established and some continue to this day.

Even the staff restaurant was not immune from practical jokes. One R/O was always very fond of his plaice and chips and the restaurant staff were well aware of this. So when it was next on the menu they fried a piece of cardboard in batter and put that on his plate. It wasn't until he cut into it that he noticed something was awry; the stifled laughter from the kitchen area gave it away somewhat. I recall he wasn't best pleased.

The restaurant manager at the time (Beryl) was affectionately known as 'Enid Whipcurd' and her penchant for cooking main courses with minced beef was well-known. When the oil tanker came to top up the tanks for station heating, it became affectionately known as the 'mince tanker' for allegedly topping up the staff restaurant mince reserves. She also took great umbrage at her 'restaurant' being called a 'staff canteen'.

The GKA football team in the early 1980s.

To be fair, the restaurant gave good quality meals at a very reasonable price; they were also responsible for stocking up the tea trolleys which came around the wings of the old station at 0930 and 1430 each day. It was intended that the trolley would come around the station console by console so that the R/O would not have to leave his or her seat. In reality, however, a queue would form at the entrance to 'A' Wing (nearest the restaurant) in order to get the pick of the salad, sausage or bacon rolls and a cup of tea or coffee which would populate the trolley. An evening trolley would also come round at 1900 to feed the late shift, which again proved very popular.

Finally, one cannot compile a history of such a well-loved establishment without remembering some of the characters who worked there. Most of the staff got on with their work quietly and efficiently, but others made a name for themselves for various reasons:

John (Jock) Bridges: Ex-Wick Radio R/O who had survived the North Atlantic and several torpedo attacks. Before the days of no-smoking, he could always be found, as he would be under his own personal cloud – of smoke. An avid smoker and he always had a couple of packets of Capstan full strength to hand. Enjoyed a couple of tots of scotch during a night shift to keep him going. Returned to Scotland on retirement.

Francis Da Costa: Hailed from Goa apparently. He worked in the Accounts section for a long time. He ran a restaurant with his wife in to Weston-super-Mare and often turned up late and worked like a man possessed to catch up with the backlog. He eventually left and the last report of him was that he had a club/restaurant in London.

Merv Lazaro: Ex-Aircraft R/O who enjoyed R/T working at GKA. He worked for Pakistan Airways. Apparently when a film called 'Bhowani Junction' was being made he had Ava Gardner and Stewart Granger as passengers and they signed his log book.

Ted Durman: Fluent in Russian and Turkish. He also rode a plastic bicycle to work which he claimed would never rust. One spring he forgot to put his clock forward to BST. Consequently, he turned up on Sunday evening an hour late for his duty and was so mortified he vowed he wouldn't forget next time the clocks changed. Come October, and sure enough he remembered - but put his clock FORWARD, thus turning up 2 hours early for his Sunday evening shift.

Chris Troy: Famous for his 'liquid lunches' on a night shift. Ran out of beverages during one particular night, climbed out of the window so as not to disturb the night overseer, drove home to replenish his stock and returned to his working point via the same route half an hour later.

Jack Todd: One of the 'old school' overseers, a gentle soul, always immaculately dressed. Famous for his shout of 'they're coming over the top' when radio conditions allowed good reception from vessels in the Pacific area. This alerted the Search Points to direct their receiving aerials to the NW-NE direction.

Bill Madden: A very dour Scottish chap. Spoke very little and you had to almost force a Good Morning out of him. He would sit quietly (most of the time) working away. At this time we were still using the old Olympic telegraph typewriters. These really had seen better days and the Engineers and RO's doing maintenance spent many hours trying to keep them operational. They had a habit of sticking or just being a pain. One morning there is a tremendous bang as Bill picks up his typewriter and bangs it on the desk in frustration. He then proceeds to stand up walk to the window – open it – then return to his desk. Lifts said typewriter up and walks to the window and throws it out.

Some years later when this solitary, but not unpleasant man who was never seen in local pubs, never socialised, was unmarried and generally kept himself to himself, died, the story goes that whosoever tidied up his affairs found, in various drawers and boxes in his house, scores of unopened pay packets. This was a time when wages could be chosen to be paid by a number of methods, the old fashioned pay packet being one of them. Bill clearly did not have a lot of faith in banks.

Arthur Lum: He was a demon with a bug key. A Japanese Hi-Mound type in a black plastic box. Always wound up to give extremely rapid dots. Not bad Morse but one would say stylish and for a first trip RO it must have been a nightmare. He was at one time RO on the passenger ship Kuala

Lumpur. He wasn't from Hong Kong as we first thought; he was once asked by a young R/O what 'Gung hay fat choy' meant, only to receive the response "How the **** should I know, I'm from Liverpool".

Arthur, after Mike Pearson, was the number two overtime king, and when the lists went up he was there with his pen. He was reputed to be able to run through the underfloor cable trunks to beat the opposition.

On another occasion, Arthur drove his motorbike into a parking space on Weston-Super-Mare seafront, whereupon he was accosted by a Brummie motorist, wielding a big spanner, who claimed that he should have had the parking place, and he then hit Arthur on the head with the spanner. Fortunately, Arthur was still wearing his helmet. Arthur grabbed his assailant by his free hand and bent it backwards and there was an ominous cracking sound indicating a broken finger. The motorist dropped his spanner and started jumping around nursing his injured hand, yelling for the police, and shouting "The bugger foo-kunged me" (sic). The police were called - however, neither party preferred charges.

When he started at GKA he drove down the M4 on the hard shoulder on his Honda 90. He was stopped by the police several times but made out he spoke no English and he was from China.

On one occasion he drove his car into the door of his garage, resulting in a dent to the door. With typical logic, he then reversed his car into the garage, shut the door, and drove his car into the door with the intention of straightening the dent. Needless to say, this plan did not have the desired result.

Charles Raleigh Mander: Actor, playwright, and R/O. He was RADA trained and allegedly shared accommodation with future DJ Pete Murray. He had several plays broadcast on the BBC and appeared in a few B movies in the 1950s. He also started a very successful Drama group within the station and they won at the BT drama festival at Edinburgh one year. He came from a very old family near Plymouth – his middle name could give a clue. The estate was very substantial. On one occasion he was showing some German students around the station (he spoke a little German apparently), describing what was happening in landline that morning. It was fairly busy, guys sending stuff on telex, Private Wire busy and incoming stuff being edited. Brian Stewart was lazing back in his chair sending on telex, with one leg propped up on one of those oval waste bins. As Charles and his party shuffled by he finished his little speech with "and this man here has a bad leg."

During one rehearsal at the station in the early 1980s, Charles was directing one of his plays for the National competitions and in the course of displaying one of the required moves, fell over a chair and knocked over all the furniture in the vicinity. Unfortunately, he broke the stem of his pipe in doing so, and as he staggered to his feet, he was heard to say (in his Lawrence Olivier-esque voice); "Ohhhh, the things one does for art...."

Charlie was on R/T on one occasion, lighting his pipe; he had his hair dressed with some sort of inflammable dressing and it caught fire – panic and confusion – however, he survived.

Jim Waldron: An ex-colonial who had worked in East Africa for many years with (possibly) the railways but as the years rolled on these men were superseded by locals and they became basically redundant. He revalidated his ticket and in later life came and joined us. He could always be found not by smoke clouds but by key clicks. He always had the widest key gap one has ever known.

Alan Padgett: Another wonderful character - always used to try to cadge a lift into work by standing on the corner of the road where he lived and try to flag down other GKA staff on their way to work. Most of us got a bit fed up with this and changed our route to avoid him....He was one of the first R/Os at GKA to use an electronic key – mains-powered and the lights in the operating wing flickered in time with his keying.

Charlie Merrilees: A lovely story about Charlie occurred when biking in for an 0800 start one winter, when it was just waste ground on the other side of Worston Lane. He always biked along what passed for a footpath between the lane and Old Highbridge Road, but on this very wet day came a right cropper and got completely soaked and plastered in mud. But hey, as he was right near the station, why not just clock in, take off your trousers to dry on the radiator and do your morning shift in your Y-fronts?

Arthur Smith: He was a drummer during the 1960s 'beat boom' and almost joined a famous band during that era - but found it more lucrative to join house bands at 'Gentlemen's Clubs' or similar. He was quite a virtuoso on the anglepoise lamp at GKA using pens and pencils as drumsticks. He emigrated to Australia but returns to the UK regularly.

Fred Bland: He seemed to spend most of his time on night shifts. He was one of a few who liked the more relaxed atmosphere of nights, plus it enabled him to take a few beverages at lunchtime with his good lady. Always sitting in 'C' Wing with his ashtray and roll-ups, usually wearing braces. His log was an endless list of ships worked. He with a few others almost vied to see who could work the most ships during a shift. Apparently he had been with Niarchos for many years.

Joe McCabe: Our artist and cartoonist. Many will remember his cartoons which featured in the Marconi Mariner and other Post Office/BT publications. Some of his cartoons are featured in this book.

Jim Sheldon: Talented but eccentric R/O from Yorkshire who used to work ships lying almost horizontally in his chair. It was bizarre - sitting way back - almost horizontal, his head barely visible above the desktop. His arms would be reaching up to the typewriter. We got used to this pose (and his other eccentric ways) over the years and it was barely commented upon. I think he got bored easily and liked to play around.

One day the station manager (Don Mulholland) brought some posh-suited visitors into the station. He showed them the Search Point position at length and then walking them up the wing a bit said: "If you come up here I will show you a work point". All the other points were manned and he had selected the only vacant place. Judging by the time spent at the Search Point, an in-depth explanation of the functions of the various switches was planned. However, it was not a vacant position.

Jim was sat there barely visible in a typical reclining pose. Just his fingers reaching up to the typewriter as he typed.

As they reached the work point, and Don realised Jim was 'lying there' in such a bizarre fashion, it caused some embarrassment. Don looked horrified and instantly decided it was not that important to show them a work point.

He marched them quickly past Jim giving a wave of his hand "those are all the work points" and through the wing saying "and up here we have our R/T room - we'll have a look in there".

You will need to picture Jim's pose for the full comedy of the incident, and also to see the horror in Don's eyes.

Jim was once observed standing up punching away at a typewriter. When he had finished someone asked him why he was typing standing up. "I was receiving a message for Her Majesty the Queen" he solemnly replied.

Tony Roskilly has a favourite memory of Jim: "I recall asking him on nights why being such an intelligent and capable man, he acted as he did when at work. His reply was classic Sheldonese: "I brought my mind to work once, it got damaged, I never did it again". I know one or two R/Os who genuinely believed that Jim's act was for real.

Pat Stevenson: Another R/O who adopted an unusual operating position. Reclining on his chair, feet on the console, he would work ships in a 'relaxed manner', much to the frustration of some of the more 'traditional' overseers.

Larry Summers: Sports-mad R/O from Liverpool who would try his hand (and normally succeed) at anything. Banned from playing football 'sine die' for attacking a referee. Had the ability to work a ship, talk about car maintenance and drink tea at the same time. On one occasion he played in a team in the local soccer league. Calamity struck one day when neither the home nor the away team remembered to actually bring a ball. However, Larry saved the day, by having his own football in the boot of his car.

Sadly partway through the game he was sent off. Being somewhat peeved, he pointed out to the ref who it was who owned the football and took it with him as he trudged off the pitch, forcing the game to be abandoned.

Don Matheson: Quietly-spoken Scotsman who never swapped duties, walked to and from work each day and refused any lifts back home in torrential rain. Unusual way of sending Morse by using his knuckles.

Vas Hira: Greek-Cypriot R/O who lived virtually opposite the station but was very rarely early for work. Most useful in assisting with the delivery of Greek-language phonograms and dealing with the numerous Greek shipping companies who used the station. Won the football pools on one memorable occasion.

Ken Wilson: Up until he left operational duties and worked permanently in the Writing Duty, the person that the duty night overseer did not want on the night shift (a good-natured wish I hope!!) was Ken. He had a wonderful store of anecdotes and if the mood took to him in the middle of the night - once the traffic rush was over - would get into his stride and that was the end of work for a while as he regaled a willing audience. Even those who won the draw were reluctant to go home if Ken was in full flow.

Ken Strachan recalls: "Resplendent with reading glasses around his neck, he used to allocate the overtime sheet and shout around the wings at short notice. When they installed the new barriers on the entrance and exit to the car park they left a space on the left-hand side. Ken was stationary at the entrance one morning with the barrier lifting when I whipped past him on my motorbike. Unaware that the pressure plate had been installed past the barrier, it came down on the roof of Ken's car. I fully expected some harsh words but none were forthcoming. However, after a month of my name absent from the overtime sheet - I think he made his point!"

Barry Taylor: Lived in a caravan on a local site and used to bring his washing in on nights to dry. Many a night shift was spent alongside his 'smalls' steaming away on the radiators. Also used the station facilities to charge his car and caravan batteries overnight. On one occasion he spilt battery acid over the floor and was subsequently banned from further charging within the station building.

Barry Williams: Welsh R/O with a penchant for working nights. Often to be working ship after ship, cigarette in hand, with a flask of hot soup by his side. Allegedly only worked the ships with the strongest signals but there is no evidence to support this. Known affectionately as the 'soup dragon'.

So as you will now be aware, the station, although highly efficient and well respected, had more than its fair share of characters and incidents. In many ways a unique and indeed enjoyable place to work. Great times.

ANNEX 1

RUGBY RADIO STATION by MALCOLM HANCOCK

The Beginning

Following the 1st World War, the Government were determined to set up the long delayed, Imperial wireless network. They wanted the station in this country to be Government-run and the General Post Office were given the job.

Hillmorton, Rugby was first mentioned as the chosen site on 14th June 1923 and on 1st January 1926 the GBR (16 kHz) Long Wave (Very Low Frequency) service opened. The transmissions were in Morse code with a transmitter output power of 350 kW - at that time the world's most powerful transmitter using thermionic valves, giving world-wide coverage. The aerial was suspended from twelve 820 ft masts built by Head Wrightson & Co. Ltd. of Thornaby-on-Tees.

In 1927, just a year after the Radio Station opened, the first radio telephone service from the UK to the USA began. Later this service could carry a maximum of two telephone calls using a frequency of 60-68 kHz in the Long Wave band. The cost of a call, during the first year of service was

£15 for three minutes, about £600 at today's prices. The service was transmitted from Rugby and the receiving station for the return leg of the circuit was at Wroughton in Wiltshire. Later a receiver at Cupar in Scotland was also used. In the USA the receiver was at Houlton in Maine and the return leg transmitter at Rocky Point, New York.

Short Waves (High Frequency)

Although circuits on Short-Wave were not quite as reliable as Long Wave, the transmitters needed much less power and smaller aerials. This reduced costs and increased demand for short-wave circuits. Therefore, a second building (Rugby 'A') was opened in 1929 and by 1935 Rugby was capable of transmitting many, mainly telephone, circuits to any country anywhere in the world.

During the 2nd World War from 1943-1945, Rugby transmitters were used to jam German night fighter W/T transmissions. This was called operation 'Corona' and used fluent German speakers at RAF Kingsdown to send out false and misleading information.

Rugby Radio at its Peak

The demand for long-distance, point-to-point, radio links continued to increase. Rugby Radio reached its zenith in 1955 when the third building opened. The size of the site was increased from 900 to 1600 acres by the purchase of land on the Northamptonshire side of the A5 Watling Street. A new building, Rugby 'B' as it was called, was opened in July 1955 by the Post Master General Dr. Charles Hill (The wartime 'Radio Doctor'). It contained twenty-eight 'Marconi' 30 kW high-frequency transmitters. Rugby now had 57 transmitters and was the biggest radio transmitting station in the world. Receiver sites were at Bearley near Stratford-on-Avon, Somerton in Somerset and Baldock in Hertfordshire.

In the mid-1960s circuits at Rugby were set up for use by NASA on the Mercury and Gemini space flights.

The 'Concorde' aircraft also had a radio circuit through Rugby. Because of its height and speed, it couldn't use satellite services. So when out of VHF range over the Mid-Atlantic, it used HF for its 'Speedbird' communications back to British Airways HQ.

During the Falklands War in 1982, a special South Atlantic short-wave circuit was urgently set up for the MoD. This, together with the GBR VLF transmitter used by the MoD(N), helped in the war effort.

The Beginning of a Slow Decline

When the first transatlantic cable opened in 1956 some of the traffic started to move away from Rugby. Over the next ten years, Rugby changed from carrying the high-density telephone traffic of North America and Australasia into providing better telephone and teleprinter services to countries in areas that had no direct cable access. These countries, in Africa, South America, Asia, Iceland, etc., only required a small number of circuits. Press services for Reuters and London Press Service (LPS) continued with teleprinter news and picture transmissions.

With the advent of satellite communications, Short-Wave International land-based, point-to-point, radio services continued a slow declined between 1975 and 1987. Other Radio Stations closed, as the remaining services were concentrated at Rugby. Between 1987 and 1992 the station was converted to carry an improved long range maritime service. Smaller modern transmitters were installed and the aerials changed. The control centre for this maritime service was at Burnham-on-Sea (known as Portishead Radio) and the receivers at Somerton.

Finally, even the ship services started to transfer to satellite. The number of transmitters required reduced and all maritime commercial services were concentrated at the 'B' Building. The 'A' Building closed as a radio station in the early 1990s.

The End is Nigh

The ship distress system moved over to satellite and ship owners, if they purchased satellite equipment for their ships, no longer had to provide Radio Officers to maintain a 24 hour Morse distress watch. Obviously, that worked out cheaper. OFTEL also relinquished the requirement that the ships service must be provided. The 'B' building maritime services finally closed at the end of April 2000. This left only two telegraphy and the Time Signal services at the original building now called Rugby 'C'.

The GBR Transmitter

As mentioned above, GBR is the call sign of the original 16 kHz VLF service that opened on 1st January 1926. The original transmitter designed and built by the General Post Office was gradually improved over the years. It suffered fire damage in 1943 but a partial rebuild and further improvements saw the original transmitter continue through to 1965. The transmitter was then completely replaced by a new three-valve, 'latent heat of steam' cooled, transmitter which was also designed and constructed by the Post Office. During and following the 2nd World War the traffic moved from commercial telegrams to ships and diplomatic news broadcasts, to Air Ministry weather forecasts and finally played an important part in the Cold War, providing submarine communications for the Royal Navy.

The MOD(N) contract for the remaining two telegraphy services, one of which was GBR, ceased on 31st March 2003 and eight of the twelve 820 ft masts were demolished on the evening of 19th June 2004.

Time Signal services

From 1927 to 1986, at certain specific times of day, time signals were transmitted on the GBR VLF service. This enabled ships anywhere in the world to synchronise their chronometers for navigation purposes.

From 1950 to 1988 time signals were also transmitted on Short Wave – 2.5, 5, & 10 MHz and for some periods on other Short Wave frequencies.

The origins of the 60 kHz time signal service (known by its call sign as MSF) go back to the 1927 Long Wave Telephone service. By 1950 Long Wave telephony had become a standby requirement and therefore the transmitter and aerial could be used for 1 hour a day for this new important time signal service. This became a 24-hour service in 1966 and was run by PO/BT on behalf of the National Physical Laboratory. A new solid-state transmitter was installed in December 1998.

The time source improved over the years from Royal Greenwich Observatory landline, to local clocks controlled by 'Essen Ring' crystal, then Rubidium and finally to a Caesium Atomic frequency standard in 1976.

The MSF 60 kHz service moved to Anthorn Radio Station near Carlisle on 1st April 2007. So, after 80 years, this brought to an end the transmission of Time Signals from Rugby.

Loran C

A temporary 'Loran C' navigation service started on 100 kHz in mid-2005. This used an American 'Megapulse' Accufix 7500 Solid State transmitter and a new 'T' aerial, straddling the building, between old 820 ft masts Nos. 1 & 2. This temporary service ceased on 4th July 2007 and was the last radio transmission from Rugby Radio Station.

The End

The remaining four 820 ft masts were demolished at 1500 BST on Thursday 2nd August 2007.

ANNEX 2

LEAFIELD RADIO STATION

In 1912 due to the geographical location, elevation and the availability of the Crown Lease, the Marconi Company on behalf of the Post Office started experimental radio transmissions on the site. Prior to this, there were already a few makeshift huts where experiments had been made of meteorological nature.

Work was interrupted during the Great War of 1914-1918 but the importance of the site was regarded by the government to warrant protection by troops. Soon after the war development resumed on what was to be the Imperial Wireless Chain. This was a scheme to provide radiotelegraph communication between countries of The British Empire using spark transmitters of 300 kW power. Its sister station was situated at Cairo and was destroyed during the Suez War in 1954.

Thirteen masts each being 305 feet high were erected. These were originally constructed of steel sections bolted together to form a hollow tube. Due to weather damage, the masts required replacement in 1943, but there was at that time a shortage of steel and it was decided to encase them in concrete to prevent further damage, increasing their weight from 50 tons to 120 tons each.

It played an important role during the war and was constantly guarded by troops and the local home guard units against possible attack. The buildings were camouflaged and a decoy building was built in plywood in the hope of confusing enemy aircraft. Leafield also provided an important long wave communications link during the Falklands War in 1982.

In 1961 a start was made on the construction of a new radio station which was built at a cost of £1,000,000. The site took on an entirely new look with large modern looking buildings surrounded by a maze of wires supported on about 80 masts of 180 feet high. The large concrete and steel masts having been demolished. The station was officially opened in 1962 and was adjudged to be one of the most modern, powerful and finest in the world. The modern STC 'QT' series transmitters were used for the maritime radio telephone service.

The radio station finally closed down in June 1986 due mainly to the advancement of satellite communication services, the remaining workload of transmissions being transferred to Rugby Radio Station.

After completion of the new radio station, a decision was made to set up a training scheme on the site for apprentice engineers. The old transmitter rooms were no longer required so these were converted for use as lecture and practical demonstration rooms.

On the closure of the radio station, it was decided that the college and radio station buildings should merge with a view to demolishing the old utility buildings which had reached their life expectancy and were becoming outdated and expensive to maintain.

In September 1986 the first phase of the conversion of the radio station building into college facilities was completed, and by the end of 1992 BT management decided to merge the training wings of BTI and BT. The Leafield BTI College was closed by September 1994, ending an era in communications history.

ANNEX 3

ONGAR RADIO STATION

The original station was built in 1920 by Marconi's Wireless Telegraphy Co. In September 1929 control passed to the Imperial and International Communication Co. when the telegraph communications of the Empire were placed in the hands of a single company. The station was transferred to Cable & Wireless Ltd in June 1934 and remained so until the passing of the Commonwealth Telegraph Act 1949 which integrated the United Kingdom radio services of the Post Office and Cable & Wireless Ltd on 1st April 1950.

The four original long wave transmitters installed in A, B and C stations provided the first telegraph services from Ongar connecting London with Paris and Bern followed by short-wave transmitters in 1924 which radiated from simple vertical aerials of no specific length.

1930 saw the introduction at Ongar of the beam aerials developed by C.S. Franklin which concentrated the energy radiated into a narrow beam, resulting in a much-improved signal-to-noise ratio at the receiving station. The beam aerials remained in use until 1966 when, due to rising maintenance costs and the demand for aerials with greater frequency flexibility, they were felled and replaced by rhombic and log periodic aerials.

In 1943, D station was constructed to accommodate five short-wave transmitters, one with remote frequency change facilities. This was followed by E station accommodating a further six transmitters. Bother these buildings were unattended, start-stop and aerial changes being effected from the main C station.

In 1959, seven remote wave-change transmitters installed at D station came into service to cater for the introduction of multi-channel telegraph systems.

1971 saw the opening of the new station, a major development towards the automation of the Post Office's international radio services. The new station, together with its sister station at Rugby, provided all long-distance HF radiotelephone, telegraph and telex circuits foreseen over the next decade or so.

The new station comprised:

Group 1: C station, providing 7 services from eight 10 kW transmitters.

Group 2: C station, providing 6 services from six 30 kW transmitters.

Group 3: D station, incorporating seven 30 kW transmitters including remote wave-change facilities and one self-tuning transmitter.

Group 4: E station, providing 7 services from eight 10 kW transmitters.

However, as satellite communications developed, the use of terrestrial services declined, meaning that many of the maritime transmitter sites were closed as part of the rationalisation process. Ongar's maritime service closed in 1985 and some transmitters relocated at the Rugby site.

ANNEX 4

DORCHESTER RADIO STATION

Dorchester Radio Station was built in 1925 and began operations in 1927, providing a global telegraph network via the Marconi Company. The station was an important part of Marconi's network of beam wireless stations.

Technology involved at Dorchester included massive beam antenna arrays with their distinctive 'T' shaped antenna masts that dominated the Dorchester skyline for many years, beaming signals to the USA, Canada, South America, the Middle and Far East.

Dorchester initially transmitted messages and telegrams to the USA and South America and as the business grew so did the number of destinations to which Dorchester transmitted.

The Middle East and Japan were soon added. Similar Marconi beam stations were built at Bodmin, Bridgwater, Grimsby, and Skegness for communications with the British Empire countries and the GPO established a number of radio stations, Rugby being the largest, to provide an international radiotelephone service.

During the Second World War securing telecommunications with allies became essential and radio stations were camouflaged in an attempt to avoid being bombed.

Post-war, Dorchester had four extra transmitters installed and communicated with South Africa and Australia, but as submarine cables and satellites started to take over the rapidly growing volume of phone and data traffic, many radio stations became redundant.

Dorchester switched operation to ship-to-shore communications and the large antenna systems were replaced with more modest antennas.

The ship-to-shore systems were eventually superseded by satellite communications and Dorchester transmitted its last message in 1979.

ANNEX 5

CRIGGION RADIO STATION

Criggion Radio Station was created as a direct result of the Admiralty's concern that Rugby's VLF transmitter (GBR), vital to the war at sea, had no standby and might be severely damaged or destroyed by stray bombs intended for nearby Coventry.

A programme was devised in 1940 to rectify this situation and to provide additional high-frequency radio transmission capabilities across the Atlantic. Criggion's VLF antenna was hung from three free-standing steel lattice towers at 182.9 metres (600 ft) tall, three guyed masts at 213.4 metres (700 ft) tall and a rock anchor.

In September 1942, the first HF transmitter was put into operation, and early in 1943 while Criggion's VLF transmitter was still in the testing stage, Rugby's GBR station caught fire. Testing at Criggion was accelerated and within just three days Criggion's new transmitter had taken over the Admiralty service to HM ships at sea. Rugby was out of action for over six months.

It was this equipment that transmitted the order to sink the Scharnhorst in 1943, it also gave the order to attack the General Belgrano on the 2nd May 1982 and to invade the Falklands.

Operating on 19.6 kHz with the call sign GBZ, the station was used until its shutdown on 1 April 2003 for sending messages to submarines. The towers and masts were demolished in August 2003.

ANNEX 6

BALDOCK RADIO STATION

Baldock Radio Station began operating in 1929. It was created as a part of the Imperial Wireless Chain, an international communications network to link the countries of the British Empire. In that same year, it helped to complete the first transatlantic radiotelephone call, made to the UK from Rocky Point, New Jersey, USA.

In 1938, Baldock Frequency Control Station opened at the same location as the main radio station. The new frequency control station played a vital role in keeping the radio spectrum clear of interference, which remains a key function of the station today.

Baldock's role changed dramatically with the onset of the Second World War. It became one of many locations which conducted the UK's wartime surveillance operations. Baldock would intercept transmissions from

German U-boats. These intercepted messages were then sent to Bletchley Park to be decoded by code-breakers.

Around this time, the station also began to go by the name of 'Slip End' – due to a nearby town of the same name, and was duplexed with Rugby Radio. In this case, Baldock was the listening station (receiving) and Rugby the sending station (transmitting). This became the core of the maritime HF radiotelephone service, with receivers at Bearley and the radiotelephone terminal at Brent being utilised to connect calls to the international telephone network, before the service was centralised at Portishead Radio in 1970.

After the war, Baldock joined the International Telecommunications (ITU), which gave its monitoring work a new global context. The data it collected would now be shared with other monitoring stations around the world, to help coordinate international responses to radio wave interference. Reports of interference to Portishead Radio's frequencies were regularly reported to Baldock for further investigation.

Ownership of the station has changed several times in its 90-year history. Five different regulators were brought together under the Communications Act 2003 to create OFCOM, who have managed Baldock ever since.

ANNEX 7

BEARLEY RADIO STATION

Bearley was a former BT/PO radio station located in Warwickshire, located on the site of the former RAF Snitterfield, approximately 5 km due north of Stratford Upon Avon. It was a receiver station, being paired primarily with Leafield transmitting station.

The site had been used as an HF radio receiving station since the early 1950s and was part of the General Post Office External Telecommunications Executive. In those early days, the receivers were Marconi GF552 units, which used valves.

A new station was built in 1967 and which was 'state of the art' with new transistorised telephony receivers, the Plessey PVR800. Each receiver occupied a full 19-inch rack standing 6 ft tall. There were no visible controls except 6 banks of thumbwheel switches to set the channel frequencies. Everything else was managed from a Central Control Unit, CCU.

The telegraphy receivers were modified Marconi HR11s which again occupied a whole 6 ft rack. The HR11s were also controlled from the CCU. The receivers were located at one end of the building and the multiplexing and decoding equipment at the other with the CCU in the centre.

The telegraphy side was mostly multiplexed teleprinters using time division multiplex, TDMX. It was possible to get up to 48 channels on a single frequency. All the decoded signals were sent down the line to various terminal buildings in London. There were a few racks of line equipment which were often visited by colleagues from PO telephones.

There was a vast aerial farm and as the station was built on a redundant World War II airfield so space was not a problem. A minor road ran through the site so there were aerials on both sides of the road and the feed lines crossed the road on high poles.

The antennas were rhombics at various heights, including a ring of rhombics on 75 ft wooden poles spaced roughly every 12 degrees of the compass.

There were also groups of 3-wire rhombics on 150 ft lattice masts and some on experimental designs ion 300 ft masts. Each was fed at both ends meaning they could be used to receive signals from either direction.

Each antenna was fed by open-wire line supported by standard telegraph poles. The wires came back to a gantry where they connected to broadband transformers and fed using gas-filled 75-ohm coaxial cable into the building via an underground duct. The cable ended up on a patch panel where any receiver could be connected to any antenna.

The station closed in 1980 when the last of the HF circuits transferred to satellite systems.

ANNEX 8

SOMERTON RADIO STATION

Somerton Radio Station was opened on 16 December 1927 following the establishment of the Marconi Beam Wireless Services, which used high quality directional receiving aerials. The first services operated through Somerton were on point-to-point routes to New York and South America, soon followed by services to Japan and Egypt. Somerton continued to expand during the 1930s and 1940s, as the importance of beam services upon the traffic receipts of the cable companies was realised, further circuits being established following the closure of beam receiving stations at Skegness and Bridgwater. During the Second World War, Somerton took over the traffic duties of the bombed Cable and Wireless office at Moorgate.

In April 1950, the station was taken over by the Post Office, and a major re-equipment programme commenced. Somerton reached its peak in the mid-1960s and took over some telephony work for the first time with the closure of stations at Brentwood and Baldock, later reverting wholly to telegraph operations. From 1976, Somerton became important as a maritime radiotelephony station, supporting the facilities at Burnham which were inadequate for the amount of traffic handled. Staff from the Burnham site were transported each day by minibus to man the R/T consoles. Following the completion of the new station at Burnham, Somerton's aerial facilities became the receiving site for the HF maritime radio service, with receivers and aerials connected to the Burnham station by microwave link via the large repeater mast on Pen Hill, near Wells.

Some of the Somerton buildings with the aerial masts in the background.

The station closed on 30th April 2000, in common with the Burnham and Rugby sites, as the HF maritime radio service ceased for ever.

ANNEX 9

BRENT RADIO TELEPHONE TERMINAL

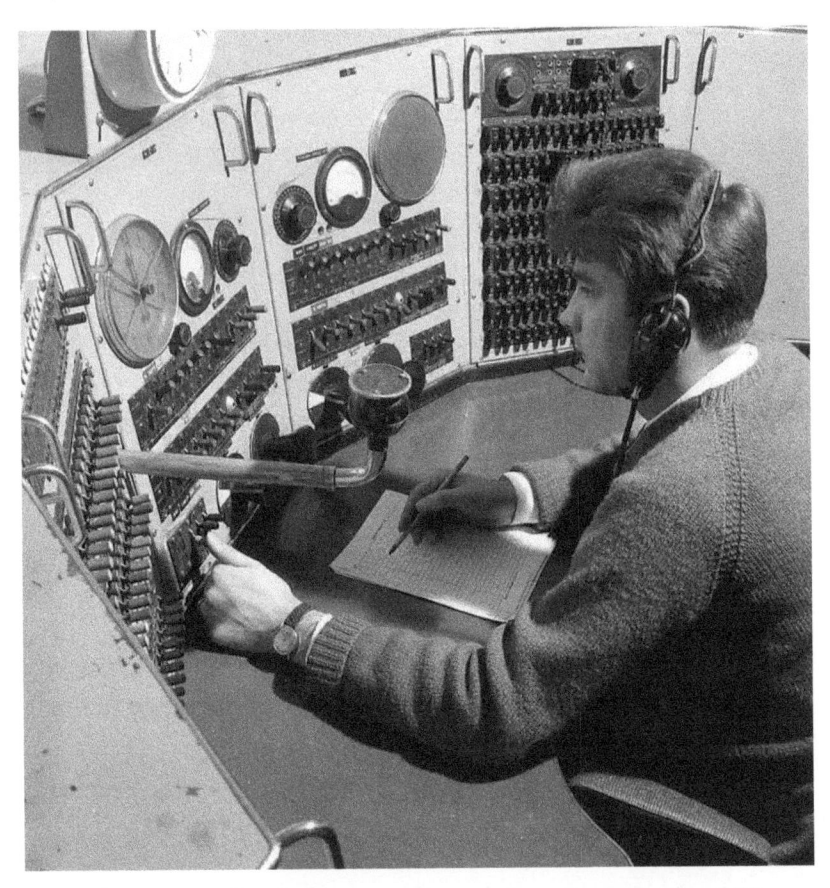

Long-distance radiotelephone calls passing through the UK were connected, controlled and timed at the International Exchange, Wood Street, London, where access to national and international trunk lines were available on a 'two-wire' manual switching basis. The radio circuits were based on a 'four-wire' basis, i.e. one pair of wires carrying signals in one direction and the other pair in the other direction. The transformation from 'two-wire' to 'four-wire' working took place at the Wood Street building, but owing to shortage of accommodation in central London, the Radio Terminal apparatus was located at Brent Building, Hendon and was the engineering control point for originally 40 radiotelephone circuits, subsequently expanded to 80 circuits.

Four principal types of equipment required for the terminal were installed in separate rooms in order to facilitate the separation of apparatus maintenance duties from those of the technical operators who control the radio circuits.

These four main divisions were:

The Operating Room: Contained operating positions at which each technical operator could control four radio circuits, and concentrator positions at which any number of circuits up to 24 could be monitored. Each operating position had a central control panel providing telephone and monitoring facilities enabling the telephone operator to listen and speak on the circuit, and volume indicators for measuring the strength of the speech signals on any of the four radio circuits.

The Radio Terminal Apparatus Room: Contained the 'Terminal Bay' which included a singing suppressor, automatic gain controlling devices, volume indicator, signalling unit and control panel. The bay could be used as a complete radio terminal although it was controlled by the Operating Room. The automatic gain control devices were provided to give a high standard of performance and enabled all but the most difficult radio circuits (such as those to and from ships) to be operated without frequent attention by technical operators.

The Privacy and Channelling Apparatus Room: Contained equipment used to ensure the privacy of calls by either a simple inversion of frequency or band splitting. As many radiotelephony calls used single sideband techniques, the Channelling equipment was used to allow another conversation to take place on the opposite side of the carrier frequency. This made it possible to accommodate two conversations on each side of the radio carrier frequency.

The Broadcast Programming Control Room: Contained an operating position for controlling and switching networks of landlines and radio circuits used in international relays of broadcast programmes and public addresses. These circuits transmitted the wider band of frequencies required for broadcast speech or music.

The Terminal worked in association with the Baldock Receiving Station and the Rugby Transmitting Station before the maritime R/T service was relocated to Burnham-on-Sea in April 1970.

ANNEX 10

THE MORSE KEYS OF PORTISHEAD RADIO

It is not known what type or model of Morse key was used at the Devizes station. Photographs of the operating area are indistinct and written information about the station sadly does not record the Morse keys in use. The first recorded evidence of the provision of Morse keys at the Highbridge site came from 1926, when PS213A models were supplied to the Post Office by Marconi. These could be identified by slotted head bearing retaining screws and three cheese head 2BA terminal screws.

Marconi PS213A key from the 1920s with the long 'tongue' clearly visible.

These continued to be used until the new station building came into operation in 1948, when new PS213 models (with Circlip bearing retainers) were obtained. These new models continued to have the long 'tongue' which tended to bend or indeed break after prolonged use.

These came with Bakelite covers which prevented the adjustment of tension and gaps, much to the annoyance of the operators. However, in 1950, one of the Rugby Engineers was tasked with the job of adapting the keys by shortening the contact strip (due to the contacts bending from consistent use) and turning the pillars around by 180 degrees. The remodeller was Gordon Bultitude, another of the engineers who before joining the Post Office had been a watchmaker and he made the prototype with long shaft and ball bearing movement. This was used by the Rugby workshops to make keys for the whole Wireless Telegraphy Section of the Post Office.

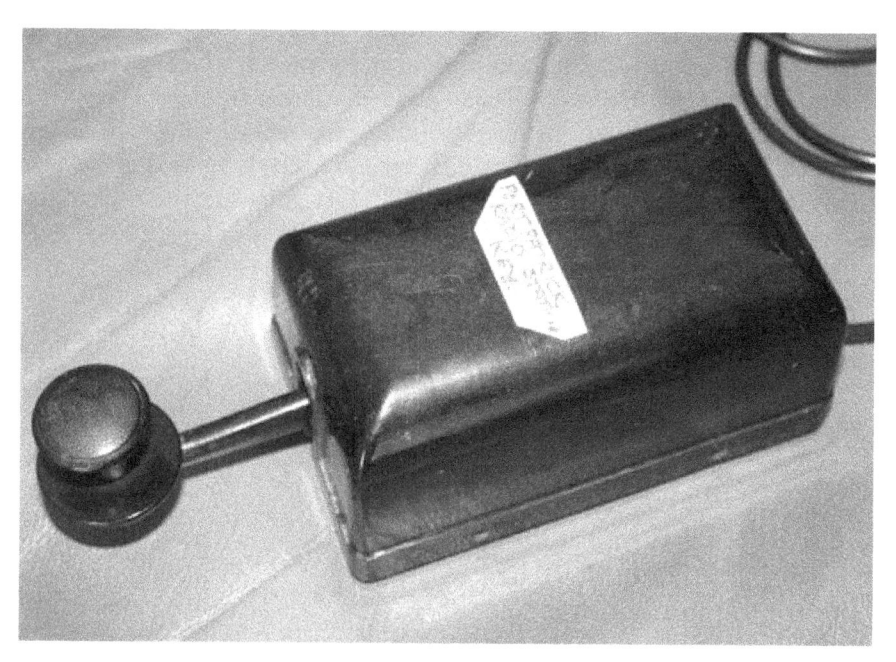

Marconi PS213 with Bakelite cover. This example is ex-Portpatrick Radio/GPK.

An example of the PS213 key as used at Portishead Radio until 1982. The 'tongue' has been shorted and a Circlip bearing retainer inserted.

In the early 1970s, a new operating wing (D Wing) was opened at the Highbridge site. There were not enough keys in stock to populate the 16 new consoles so new Norwegian Elektrisk Bureau keys were purchased, exclusively used in that wing. These divided opinion amongst Portishead Radio staff, some finding them a bit lightweight and 'springy', although they were perfectly adequate for professional use. The keys used in 'A' and 'C' Wings continued to be the modified PS213, which were also used at the broadcast position in 'B' Wing.

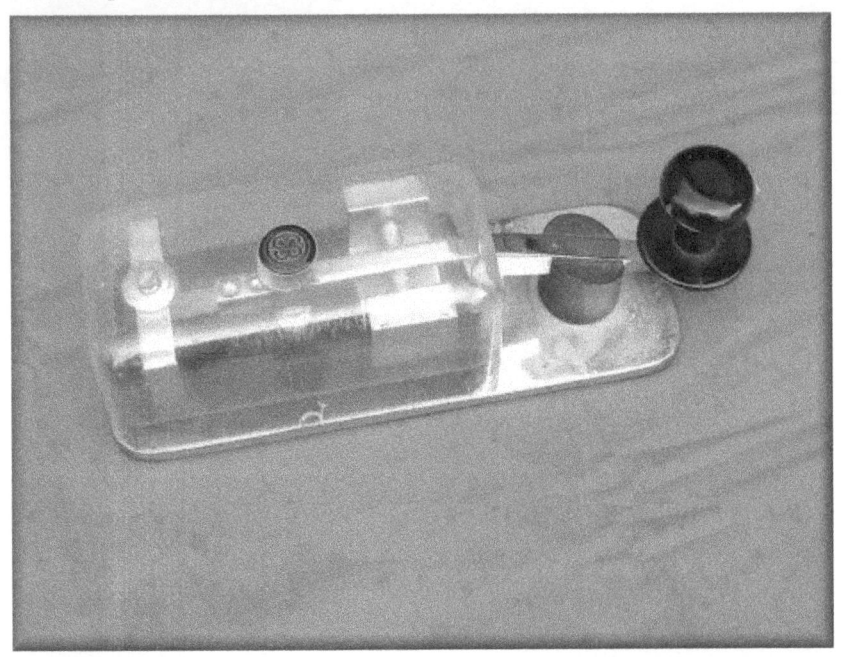

Elektrisk Bureau Morse Key as used in 'D' Wing 1971-1982.

When the station moved to the new building in 1983, the Rugby Central Workshops manufactured keys based on the Marconi PS213A model. These featured 3 smaller knurled head terminal nuts, rather than the large 2BA cheese head terminals of the earlier keys. These were mounted on heavyweight metal bases with rubber grips on the base to ensure they would not move around on the new consoles whilst being used.

A label was added on the base stating that they were manufactured by 'Central Workshops Radio Station Rugby'. Only 80 of these keys were ever made, and were of high quality; a dream to use.

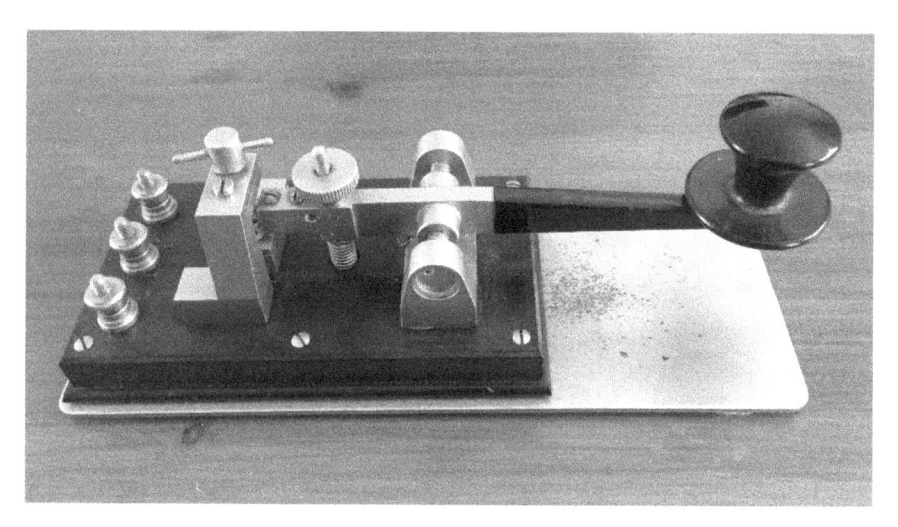

The 'Rugby' Key.

For the first time, electronic keys were provided on each console. Previously, staff who wished to use their own keys had to connect them by the judicious use of crocodile clips to the contacts on the standard key mounted on each console. One enterprising R/O brought in his own mains-powered key which had the unfortunate tendency of flickering the lights in the operating area in time with his Morse.

In the early days of automatic and electronic keys, staff were requested to undertake a test to prove their competence in using such keys. The tests were based on the standard proficiency test at the time, which involved sending text at 27 words per minute with only 4 corrected errors and no uncorrected errors. Plain language, code groups, figures, punctuated characters, and accented characters were all part of the examination.

These tests were, however, lapsed when virtually all staff preferred to use electronic or mechanical (Vibroplex) keys, and their competence in using them was beyond question.

When it came to selecting the preferred electronic key for the new building, staff were heavily involved in the selection process. Three candidates were selected from the more popular and recommended keys on the market at that time. They were subsequently evaluated, and comments from staff members were recorded in a notebook for examination at the end of the trial.

The keys tested were:

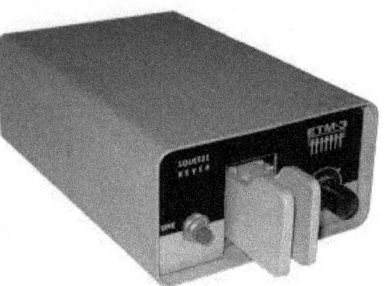

Katsumi EK-150

Samson ETM-3C

Heathkit HD-1410

After a few weeks of constant testing, with the Heathkit Key withdrawn from the final evaluation due to reliability issues, the Katsumi EK-150 was selected. However, many R/Os continued to use their own keys (Vibroplex models being particularly popular), and some staff actually purchased their own Samson ETM-3C key for their personal use.

Each EK-150 key was modified by local engineers to allow the power supply and signalling line to be combined into one plug and socket to enable easy transference between consoles.

One staff member made his own key based on the G3KHZ design described in amateur radio magazines of the time, and sold them to staff members. These proved popular and many staff used them on a regular basis.

All the surplus keys from the old station were sold to staff prior to the move to the new building. However, when the station closed in 2000, most of the keys were removed by engineers and some were given as mementos to staff members leaving the service. It is not known how many keys were 'donated' to staff members as no records were ever kept of their fate.

The G3KHZ keyer, popular amongst Portishead Radio staff.

However, I am sure that some R/Os managed to obtain their own 'souvenir' keys using various means. Most of the electronic keys sadly ended up in the skip when the station was dismantled shortly after closure.

On occasions, such keys appear on online auction sites such as eBay and sell for prices exceeding £1,000 each, and are highly-prized on the radio amateur market.

ANNEX 11

OFFICERS IN CHARGE – BURNHAM (PORTISHEAD) RADIO

1925-1934	E. F. GREENLAND
1934-1941	A. F. THOMSON
1941-1945	G. F. STEER
1945-1948	W. SWANSON
1948-1950	F. A. SHARRATT
1950-1960	L. G. FROUD
1960-1962	R. M. GIBSON
1962-1972	T. N. CARTER
1972-1984	D. J. MULHOLLAND
1984-1985	A. C. HAMBLIN
1985-1993	E. W. CROSKELL
1993-1998	R. A. STEVENS
1998-1999	P. A. BOAST
1999-2000	M. J. DAVIES

INDEX OF REFERENCED RADIO STATIONS
(Devizes and Portishead not indexed as they feature throughout)

Aberdeen	15, 16, 19, 62
Alexandria	70, 72, 80
Amagansett	54
Anglesey	118, 147, 164
Asmara	72
Athens	221
Awarua	122, 133
Baldock	39, 40, 43, 46, 85, 116, 119, 127, 128, 133, 134, 141, 151, 289, 300, 301, 304, 307
Bathurst	72
Bearley	39, 289, 301, 302
Belfast	62
Bermuda	68, 80
Bern	237, 238, 255, 280, 294
Birmingham	62
Bletchley	301
Bodmin	296
Bolt Head	11, 14
Bombay	122
Bracknell	186, 189, 190, 214
Brent	127-129, 141, 301, 306
Bridgwater	296, 304
Bristol	62
Browhead	9, 165
Buenos Aires	139
Burnham/GRL	33, 62, 84, 89, 90, 100, 102, 158, 262
Caister	9, 11, 13, 33

Cambridge	62
Cape Town	122, 139, 150
Cardiff	62
Carlisle	62
Ceylon	92
Chatham	107
Colombo	80
Colwyn Bay	62
Criggion	94, 119, 298
Crookhaven	9, 11, 14
Cullercoats	13, 62, 111, 213, 228
Dorchester	19, 20, 151, 159, 170, 296, 297
Dover	15
Dundee	62
Dunstable	101, 105
Durban	139
Edinburgh	62
Elbe-Weser	280
Exeter	62
Fairseat	139
Falklands	80, 187
Gibraltar	80
Glasgow	62
Gloucester	62
Grimsby	13, 15, 33, 296
Halifax	123
Harrogate	62
Hereford	62
Holyhead	9
Hong Kong	99, 123, 127, 251

Hull	62
Humber	33, 62, 111, 129, 256
Hunstanton	13
Ilfracombe	110, 111, 140, 161, 162, 164, 205
Inishtrahull	9
Inverness	62
Irirangi	122
Kaliningrad	254
Kiel	79, 280
Lagos	72
Land's End	15, 22, 79, 107, 111, 118, 139, 140, 164, 204, 205
Leafield	14, 19, 28, 43, 159, 165, 170, 206, 292, 293, 302
Leeds	62
Lisbon	72, 254
Liverpool	9, 11
Lizard	9, 11-13
Lyngby	158, 222, 280
Mablethorpe	13, 33
Malin Head	9, 11
Malta	80, 122, 127
Manchester	62
Mauritius	122, 130, 150
Monaco	254
Niton	11, 62, 111, 118, 140, 212, 213, 228
Nottingham	62
Norddeich	79, 135, 136, 197, 280
North Foreland	9, 11, 62, 111, 118, 139, 162, 164, 258
Oban	110
Ongar	159, 170, 206, 294, 295
Oostende	280

Perth	139, 275
Portpatrick	13, 15, 20, 23, 62, 111, 164, 213, 309
Rame Head	15
Reading	62
Rogaland	135, 136, 137, 158, 252, 280
Rosslare	9, 11, 13, 33
Rugby	19, 28, 29, 39, 40, 43, 46, 47, 49, 60, 69, 70, 119, 133, 134, 151, 165, 170, 188, 199, 206, 207, 222, 223, 226, 240, 241, 2767, 288-291, 293, 295, 296, 298, 301, 305, 307, 308, 310, 311
St. Catharines	9
St. Lys	245, 280
San Francisco	157
Scheveningen	135-137, 276, 280
Seaforth	11, 33, 62
Shrewsbury	62
Simonstown	80
Singapore	123, 150
Skegness	13, 296, 304
Slidell	107
Stockholm	237
Somerton	19, 142, 144, 145, 150, 151, 159, 171, 187, 193, 199, 216, 223, 240, 252, 253, 289, 290, 304, 305
Southampton	62
Stonehaven	19, 58, 110, 111, 129, 164, 205, 263
Sydney	69, 122, 139
Tripoli	72
Trinidad	68
Tunbridge Wells	62
Valentia	14, 51

Vancouver	122, 133
Vizagapatnam	122
Wellington	122, 133, 139
Wilhelmshaven	79
Withernsea	19
Wick	13-15, 111, 129, 281

TABLE OF ABBREVIATIONS

AFC	Automatic Frequency Control
ALRS	Admiralty List of Radio Signals
AMVER	Automated Mutual-assistance Vessel Rescue
BLA	British Liberation Army
BAMS	British and Allied Merchant Shipping
BART	Burnham Automatic Radio Telex
BMHS	Burnham Message Handling System
BRRAWC	Burnham Radio Recreational And Welfare Committee
BTI	British Telecom International
CAA	Civil Aviation Authority
COMMCEN	Communications Centre
CTO	Central Telegraph Office
CW	Continuous Wave
DF	Direction Finding
DOC	Distributed Organisational Control
DTI	Department of Trade and Industry
ETA	Estimated Time of Arrival
ETE	External Communications Executive
FCC	Federal Communications Commission
GCHQ	Government Communications Head Quarters
GM	Good Morning
GMDSS	Global Maritime Distress and Safety System
GPO	General Post Office
GTL	Greetings Telex Letter
HF	High Frequency
ICSC	International Customer Service Centre
ITU	International Telecommunications Union
LF	Low Frequency

LP	Long Path
LPS	London Press Service
LSB	Lower Side Band
MF	Medium Frequency
MO	Maintenance Officer
MOD	Ministry Of Defence
MSF	Médecins Sans Frontieres
OFCOM	Office of Communications
OM	Old Man
OTF	Optimum Transmission (or Traffic) Frequency
OWS	Ocean Weather Service
PMG	Post Master General
POEU	Post Office Engineering Union
REOU	Radio and Electronic Officers Union
RLP	Rotating Log Periodic
RSGB	Radio Society of Great Britain
RT	Radio Telephony
RTL	Radio Telex Letter
S.G.	Specific Gravity
SITOR	Simplex Teleprinter Over Radio
SLT	Ship Letter Telegram
SNF	Ships Name File
SOE	Special Operations Executive
SP	Short Path
STD	Standard Trunk Dialling
TAS	Teleprinter Automatic Switching
TMS	Telegraph Manual Switching
TOR	Telex Over Radio
TRC	Telegram Retransmission Centre
UCW	Union of Communication Workers

UDC	Urban District Council
UHF	Ultra High Frequency
UPW	Union of Post Office Workers
USB	Upper Side Band
VHF	Very High Frequency
VLF	Very Low Frequency
WPM	Words Per Minute
WTC	Wireless Telegraphy Coordinator or Control
WTS	Wireless Telegraphy Section